C0000 009 284 656

Life from an RNA World

Life from an RNA World

THE ANCESTOR WITHIN

MICHAEL YARUS

HARVARD UNIVERSITY PRESS
Cambridge, Massachusetts
London, England
2010

Printed in the United States of America

Library of Congress Cataloging-in-Publication Data

Yarus, Michael, 1940–
Life from an RNA world : the ancestor within / Michael Yarus.
p. cm.
Includes bibliographical references and index.
ISBN 978-0-674-05075-4 (alk. paper)
1. RNA. 2. Molecular biology. I. Title.
QP623.Y367 2010
572.8′8—dc22 2009044011

Contents

Introduction to Your Ancestor

Therefore I should infer from analogy that probably all the organic beings which have ever lived on this earth have descended from some one primordial form, into which life was first breathed.

—Charles Darwin, *On the Origin of Species* (1859)

Remove literary, grammatical and syntactical inhibition.

—Jack Kerouac, "Belief and Technique for Modern Prose"

Greetings to you, who look with good-spirited curiosity at this congenial anachronism. It is unclear how much longer people will write on dried and flattened wood. Trees do so much for humans and for our planet that it hardly seems fair to ask them to carry our thoughts as well. But fair or not, archaic or not, it still appears oddly plausible to make a book in order to collect a substantial number of thoughts in one container. Having made one, I am thankful that it yet seems plausible to you to look at the result. Therefore, a heartfelt welcome. Here's hoping that wooden books survive the ascent of digital text.

This particular book was provoked by an advance in biological thinking about life on Earth. Its subject is a small slice of the action in Darwin's breathtaking summary above. A majority of evolutionary biologists believe that we now can envision our biological predecessors on this planet, though we have never seen them. *Life from an RNA World* is about these vanished old ones, sketching them at a long-ago time just as their workings began to resemble closely our own. What was the difference between our early relatives and their later offspring—us?

Sketching such a portrait takes some effort, for we will end up in quest of nearly the first among the living beings of our planet. Nearly, but not quite the first. So we must see our quarry across the billions of years between their heyday and ours. The effort that brings them into focus has been widely talked about within the field of biology—but surprisingly little of the story is known by those outside the laboratories who might also be interested. So here is an album to introduce interested non-biologists to our relatives in deep time—slouching along between the origin of the first rudimentary life on Earth and the appearance of more complex beings—who had nearly mastered the intricate informational handicrafts that make modern cells. This era between is called the RNA world.

It is endlessly interesting to inquire into our ancestors and to try to guess in what sense that Ukrainian or Lebanese great grandparent left us his or her gifts. But while this book is concerned with our personal genealogies, it is also about something deeper. If you step back far enough, your genealogy merges into the history of life on Earth—indeed, the only life we know. And within the RNA world lies the solution to a major mystery about the path life has taken on Earth.

To initiate that billion-year view, we need to excavate for some foundations. In particular we need some notions of what evolution is, and of what life itself is, so we can coherently speak of its youth. An image of the small, whirling galaxy of our genes will help us to appreciate the extent to which we are still the children of our great . . . great grand-ancestors. We will spend a lot of time, then, tracing the shadows of an earlier RNA world within our own cells. Such preoccupation is appropriate because we now flourish by wielding ancient, borrowed genetic recipes. Indeed it seems, surprisingly, that here in the early days of the twenty-first century we are still at the beginning of appreciating our common ancient RNA patrimony.

I am a professional biologist, doing research and teaching in areas that sometimes overlap the subject of this book. This means that later on in the book, when there is not much to go on, you will hear my opinions. I will try to state this sufficiently plainly to make these sections clearly distinguishable from those setting forth concepts that rest on concrete and wide-ranging support and consensus.

Many thanks to those who read drafts of the manuscript and made suggestions—John Abelson, Richard Byyny, Tom Cech, Nataliya Chumachenko, Shelley Copley, James Dahlberg, Matt DeYarus, Larry Gold, Teresa Janas, Leslie Leinwand, Elsebet Lund, Irene Majerfeld, Bill McClain, Peter "ribosaur" Moore, Norm Pace, Alyson Yarus—and to the indefatigable students of MCDB 4100, "The RNA World." Particular thanks to all those who made known their views about good and bad words, and to the artist who helped me picture these thoughts, Greg Kuebler. Many thanks also to the Graduate School of the University of Colorado, whose Council on Creative Work gave me a year free of other responsibilities to work on this book.

I also explicitly acknowledge those in my Boulder laboratory who actually did the work summarized in Chapter 16 on the reactions of translation. I do this because I wish to credit them, and also to emphasize that experiments not occur without the labors of real people, who often give years to an investigational campaign. They are as follows: *amino acid activation* —Dr. Krishna Kumar; *aminoacyl-RNA synthesis*—Dr. Nataliya Chumachenko, Dr. Mali Illangasekare, Dr. Oleg Kovalchuke, and Rebecca Turk; *peptidyl transferase reaction*—Dr. Mark Welch; *amino acid binding to RNA and genetic coding*—Dr. Shankar Changayil, Dr. Greg Connell, Dr. Mali Illangasekare, Dr. Michal Legiewicz, Dr. Cathy Lozupone, Dr. Irene Majerfeld, and Dr. Shawn Zinnen.

The book you hold is much better for these many good-spirited contributions from others. The text gathers ideas from many sources, and I am certain that I have not explicitly remembered everyone who expressed an essential thought. Accordingly, there will be those whose work changed my mind and illuminated my path, but whose exertions will not be expressly described. To these numerous unnamed thinkers and experimentalists, my profound thanks and sincere apologies, humbly offered. Nevertheless, bookmaking requires that the opinionated, crotchety author reject and accept ideas as he goes. All remaining errors within these pages are mine—sometimes achieved despite the best advice imaginable.

Before We Begin: A Voluntary Chapter

Outside of a dog, a book is man's best friend. Inside of a dog, it's too dark to read.

—Groucho Marx

This is a book for people who like books about science and are accustomed to reading them. It also intends to be a book that could be consulted by undergraduate and graduate students of life and its earthly history. Professionals in other scientific areas might also use this book to find a handle on RNA biology. You will note a ringing omission from this list of potential addressees. *Life from an RNA World* is not a book written for professional biologists, though some might read it because they take an inclusive pleasure in its topic. So, while writing, a professional attitude toward scientific citations was among the first things I discarded. Here you will not find the inclusive point of view appropriate to a textbook or a professional journal. It seemed particularly pointless to recommend a comprehensive but insurmountable heap of further reading, because a wall of footnotes gives precisely the wrong impression. Conspicuously looming reading suggests that you cannot know more without a lot of poring. In fact, the reverse is much more often true: you will know ten times more (and have ten new questions) if you read one extra, small, apt thing.

Accordingly, my reading recommendations are directed to those looking for a trailhead at the edge of the wilderness of the scientific literature. Instead of comprehensive references, I suggest a few readings at the end of each chapter, along with some explanation about why each was selected. These should be immediately informative, but they will also include their own references. Those who read on will find the biology literature swiftly opened before them. In many cases, the names of people associated with particular views, theories, or jokes are given in the text. These convey necessary references, homages, and appreciations, but they can do much more. If you look up any person named in this book in databases like Google Scholar or the National Library of Medicine's PubMed, you will launch a new reading expedition custom-crafted to suit your own curiosity.

Thus the literature trail signposted in these pages branches deeply, but just out of sight. I particularly want to encourage individual pursuit of topics that spark a flash of interest, even when the follow-up requires probing into original scientific articles. You should not be intimidated by the professional writings of biologists, or of any scientist. The barrier to entry is usually only mastery of insider jargon, nothing more rare or demanding. In approaching a scientific paper, start by reading the abstract or summary and the introduction, both of which are commonly written in near-English. Then progress to the discussion, and finally to the results, in the middle of the paper. Take the results last; they often make up the most abstruse bit of a paper because of their detailed relation to particular tricks of the trade. In navigating this elementary but intricate terrain, a good general molecular biology textbook will be a help. You might keep one, like *Molecular Biology of the Cell* or *Molecular Biology of the Gene,* close by. Remember also the lexicon at the end of this book, which attempts to outline a basic RNA vocabulary for the reader.

But keep all this in perspective: if you encounter something incomprehensible while reading, pass it by in utter serenity. Bear in mind that experts say many indecipherable things. It is their job. Making sense of things in your own terms is the only useful goal. Because some references are books, feel free to read only those parts relevant to your current questions and interests. Your reward, should you care to claim it, will be a conversational acquaintance with this kind of science—at least as conversational as the formalities of a written article allow. A scientific journal article will be more spontaneous than any textbook, but less vivid than conversation with the person who wrote it. However, I hope none of this appears to be Required Reading: I have tried to make this book reasonably comprehensible without external expeditions.

Then again, if you would prefer to encounter the RNA world in surroundings replete with a professional outlook toward scientific literature and all its trappings, you can and should read *The RNA World.* This is the leading professional anthology in the field—and I have no financial interest in it.

Readings

Molecular Biology of the Cell, Fourth Edition. Bruce Alberts, Alexander Johnson, Julian Lewis, Martin Raff, Keith Roberts, and Peter Walter. Garland, New York (2002).
This useful book is freely accessible online at http://www.ncbi.nlm.nih.gov/books/.

The RNA World, Third Edition. Raymond Gesteland, Thomas Cech, and John Atkins, eds. Cold Spring Harbor Laboratory Press, Cold Spring Harbor, N.Y. (2006).
A collection of authoritative essays by workers in the areas tangent to the RNA world, written for professionals who want a glance into the topic. The first (1993) and second (1999) editions are also available, and they lack only the recent updates for most articles. A fourth edition is on the way in 2010.

Molecular Biology of the Gene, Fifth Edition. James D. Watson, Tania A. Baker, Stephen P. Bell, Alexander Gann, Michael Levine, and Richard Losick. Benjamin Cummings, Upper Saddle River, N.J. (2004).
The latest edition of a classic text; earlier editions may serve almost as well as a first resort on basic questions of molecular biology.

Framing the Problem: The Buffalo and the Bacterium

Go on, I tell you. You have the stomach for it!

—Franz Liszt, to Edvard Grieg

Life on Earth immediately presents us with the striking constancy of individual descent. Wombats, with few exceptions, give birth to exemplary wombats. Nothing could be more obvious. All the same, this humdrum reflection presents a vast impasse to thoughtful examination: how is the plentiful detail of every creature recorded and accurately replayed in its offspring?

Alongside this constancy of tiny details is the contradictory reality of pervasive genetic change across vast time. Where did all those varied creatures come from? A protobovine ancestor becomes both water buffalo and miniature Holsteins in the long run. Why a dingo *and* a fox? Wolves have become both Chihuahua and Shar-Pei, and this last divergence has happened almost within living human memory. How can each dizzyingly complex being be successfully altered by its residence in the world? When a hesitant Charles Darwin (at his ease in England) and a fever-wracked Alfred Russel Wallace (during an illness in the Indonesian Moluccan Islands, but thinking on a peculiarly parallel track) put forward their ideas about the destiny of biological types in 1858–1859, what were they talking about? We need to agree about this now, because later on we will often talk about these ideas, and even more often assume them.

Recording and propagation first, then. We are now quite clear that each creature describes itself in an essentially linear digital recipe, which is broken into groups called chromosomes and stored inside the membrane-bound, microscopic, subcellular compartment called a cell nucleus. The text of the recipe is written in four characters, called nucleotides, and these are strung together to make long linear texts in the molecule DNA, which is coiled tightly within the chromosomes. The chromosomes consist of many linked individual recipes called genes, each describing one or a few slightly varied products. Closely linked groups of genes may be related entries, like the books shelved near each other in a library. Or, more often, neighboring genes in a chromosome may have no immediately obvious relation to each other, like books at a garage sale. These chromosomes taken altogether are called the genome, the collective name for the genes. There may be other forms of inheritance not embodied directly in DNA sequences, but those contributions to the genome are probably tiny in comparison to that written in chromosomes.

As a result, genetic texts written as chains of nucleotides—symbolized by the letters C, A, G, and T—are more monotonous than hexadecimal computer text, which can use 16 different characters for each position. And genes are much more monotonous than English text with its 26 letters. But with three billion genetic characters in the genome of mammals like us, a lot of instructions can be encoded as strings of this four-character text (compare Chapter 8).

Genomic texts are a recipe for the creation of a creature. Genetic instructions for creature assembly can be roughly translated into words: "Make *this much* of *that* gene product and put it *there* at *this time.*" Crucially, the idea of a digital creature recipe includes the idea of a change in one or more characters of the string—a mutation. That is, writing now in nucleotide text, "TAG A CAT" could mutate to become "GAG A CAT." Mutations like these change what gene products are made, how much of each is made, and where and when they are formed. They potentially yield a more or less functional organism. If, for example, you change a protein that binds oxygen in the blood, you may have a deleterious mutation, recognizable as the genetic disease called anemia. If you lose body hair, the change might be innocuous or even helpful—if you are an Olympic swimmer. Whether mutations are destructive, helpful, or somewhere in between is determined by the process called natural selection.

Darwin's answer to the second question, dealing with divergence into new creatures, dominates biology. It is still shocking—yes, shocking—to encounter his exceptionally simple explanation for the way inevitable mutations in the genetic text are sorted out, and therefore, ultimately, his explanation for the stunning variety that is life on the Earth. And in writing "exceptionally simple," I am not minimizing Darwin's accomplishment. Philosopher Daniel Dennett was not polishing his hyperbole when he wrote that Darwinian evolution is the greatest idea anyone has ever had—way ahead of relativity and the germ theory of disease, for example. There is no question that agreement about Darwin's meaning will be worth our time.

Creatures' genes vary. Creatures expressing more effective genetic texts leave more descendants, and therefore genes of their type accumulate.

This statement and its consequences, are, astonishingly, all there is to it. I don't wish to take the "consequences" too lightly; for example, the question of when enough change has accumulated to declare a new species is itself subtle and weighty. But, separated from the political and religious uproar it has engendered, Darwin's fundamental idea seems plausible, uncontroversial, so simple and convincing that it almost demands agreement. The essence of it is an exceedingly elementary bit of arithmetic: that which makes more of itself is therefore more numerous. What part of this notion seems dubious? Even some of its furious opponents concede that it must be true at some level (see Chapter 7).

Most remarkably, Darwin saw his answer clearly long before genetics or genetic texts were understood. Though he did not yet have these details, Darwin understood genetic (heritable or inheritable) variation. He knew that desirable differences like strength among horses, good humor among dogs, and abundance in crops could be captured by simple means—choosing the desirable individuals for breeding. In time the creature under selection changes permanently as its selected genetic constitution changes. Wolves, over time, become Chihuahuas as well as Shar-Peis. For natural selection (or artificial selection), what is reliably chosen is the ability to create more like yourself and, we would now say, more copies of your genes.

New species usually require some kind of isolation so that interbreeding is depressed, because the two clans are different when the isolated group again meets the ancestral population. Other routes to a new species exist, but if genetically separated groups interbreed they will not remain isolated as separate species. The essential points are descent with variation (genes and mutations) followed by selection (survival, reproduction, success of progeny). This supposes undirected mutation, followed by selection that is directional, though it has no director. Darwinian evolution is not a random process, as is sometimes charged (often by those seeking to make it difficult to understand), but rather a combination of accessible mutations (e.g., restricted by the genes we start with), followed by a selection for utility, incorporating the nonrandom mandates of the environment.

Natural selection must work in every generation, benefiting creatures of every era, because natural selection measures only today's necessities. It has no possible way to know the future. Natural selection therefore can find only the immediate good —it is ignorant of ultimate success. Darwinism is a prescription for only local and immediate achievement, therefore completely congenial to the go-getters among us. Or, if you object to the economic tinge of this expression, evolution means that change is not just an incidental characteristic of biology. Instead, living things are the visible result of a prolonged, unceasing, unsighted search for immediately useful gambits.

Darwin deduced that descent with modification logically allows there to be one ancestor of all life on Earth. This stunning notion is abundantly confirmed by every organism's possession of the same genetic code (see Chapter 18) and every organism's connection to one and the same tree of life (see Chapter 3).

But since Darwin, during the twentieth and twenty-first centuries we have considerably refined the idea of mutation. We know that a great flood of mutations flows through genomes, for example, from natural radioactivity and errors in DNA replication and repair. Not all such mutations have effects on the organism. Many genetic changes do not change either the time, the type, or the amount of the output of the genes. These are called neutral mutations—a major source of change in genomes, but not usually contributors to the variation that is acted on by natural selection. Neutral mutations accumulate (the changes are collectively called genetic drift), and they are important in any census of genome change (such as in studies of evolutionary trees or descent, as described in Chapter 3), but they are mere bystanders at the Darwinian concert, their presence as arbitrary as the hiss from a detuned radio. That arbitrary chemical and physical alteration in genomes is one of the significant differences between us and our ancestors may seem odd at first, but it is true.

Two other types of mutational sequence changes in the genetic text are selected. One type of selected mutation is the deleterious change. Deleterious changes are probably more numerous than potentially favorable changes, those that the adaptive Darwinian mill uses to make an organism better fit its world. Most likely an undirected change in a complex system will mean that the system will not work as well as the original, which had many parts selected precisely because they functioned well in their existing forms. If you hit your Stradivarius with a hammer, you are unlikely to improve its tone. Deleterious mutations are selected against and fade from a population (are less abundant in descendants) because they impair reproduction, only to potentially recur in the future because the mutational process is blind and enduring.

A second, smaller class of selected mutations is the favorable ones, rare but crucial in their impact. Thus, it is wrong to think of mutation as completely random in the Darwinian world view; many mutations do not play (the neutrals), many are eliminated (the deleterious), and only a minority finally end up persisting and changing the evolutionary fate of an organism. This minority status has the curious effect of making the adaptive mutations a small, special class not necessarily representative of total change in the genome. Thus, a significant storm of change surrounds genomes (discussed further in Chapter 4), but only neutral changes, and a selected, adaptive trickle descend into deep time.

Evolution thus has a cost, which must be visible. Because the majority must usually be less fit than the selected few whenever there is evolutionary advance, evolutionary advance implies the waning of the creatures who do not succeed most brilliantly. For a gene indeed, evolutionary success is defined only in the context of the lesser reproductive success of many other genes (and their bearers) who do less well and decline into history. Darwinian deficiencies are among the less appreciated marks of evolution—but here lies a large fraction of human art. Romeo and Juliet; Frankie and Johnnie; sex, drugs, and rock and roll—a Darwinian torrent flows onward, and it will not end until humans end.

Because mutations are usually small alterations of a huge, divided, linear digital message, progenitors and descendants usually resemble each other pretty closely. Evolution therefore advances almost continuously, rather than by broad or general change between successive generations. This is the principle of continuity, which I first heard formalized by Leslie Orgel of the Scripps Institute, and it will be useful later in this book.

And thus were formed the orchid and the spider, the buffalo and the bacterium. From arbitrary changes in the genetic text, within constraints set by the chemistry and physics of genomes, mindless but purposeful selection from this myriad, and then—forms most varied, sublime, and pertinent to the worlds that bear them.

Readings

A Farewell to Alms: A Brief Economic History of the World. Gregory Clark. Princeton University Press, Princeton, N.J. (2007).
A consideration of human Darwinian genetic change as a possible route upward from a long-stable stone age culture, providing a changed worker who could participate in and profit from the Industrial Revolution.

Darwin's Dangerous Idea: Evolution and the Meanings of Life. Daniel C. Dennett. Simon and Schuster, New York (1996).
An exceedingly rare combination of vigor and rigor, in service of the idea and explanatory power of Darwinian evolution.

The Moral Animal: Why We Are the Way We Are: The New Science of Evolutionary Psychology. Robert Wright. Vintage, New York (1995).
A Darwinian take on the possible evolutionary origins of human behavior and psychology—human nature, in short.

3

The Big Tree: No Jackalopes Please

The fruit of the righteous is a tree of life.

—Proverbs 11:30

The framework of bones being the same in the hand of a man, wing of a bat, fin of the porpoise, and leg of the horse,—the same number of vertebrae forming the neck of the giraffe and of the elephant,—and innumerable other such facts, at once explain themselves on the theory of descent with slow and slight successive modifications.

—Charles Darwin, *On the Origin of Species* (1859)

As I sit writing these words, birds glide in to drink from a pool nearby, insects buzz, and a short distance away evergreen treetops sway as the sun stirs the atmosphere. Look anywhere on the temperate or tropical Earth—you cannot help but be struck by the riotous success of life. And this despite its invisibility; life's successes are mostly unseen because virtually everywhere microbial cells outnumber the cells of visible beings. Notably, this is true even within the bodies of the visible beings—our own cells are outnumbered by the cells of microbes on and in us. A huge, usually underappreciated world of small, diverse creatures vibrates with activity below the resolution of unaided human vision. Yet there is still a simple way to picture all life on Earth together, in impartial and proper array, and it is our subject here: the Big Tree.

The Big Tree appears at this point because, in order to locate the RNA world (the era of RNA creatures) in relation to life today, it is vital to visualize the complete course of life on Earth. So in this chapter I describe how we can draw a rational, objective picture which lays out life's history, using information that is subtle, quantitative, and now within easy reach.

The key idea is this: each creature preserves a record of its history, and particularly of its relation to its nearest relatives in deep time, in its genome. That is, each living thing is most similar to the now-separate type of creature that most recently shared a common ancestor. Humans are similar to chimps, but less similar to dogs. Each is less similar to the lemur from which it diverged next most long ago, and so on. Based on this concept, and given a simple measure of genomic similarity (or conversely, evolutionary distance), we can draw a map.

The implied map is no more complex than that of the Earth's surface. Suppose that LizCity is 5 miles from Dogtown and 6 miles away from Flyville. If you know that it is 8 miles from Dogtown to Flyville, you can draw a simple map that relates the cities, like this one:

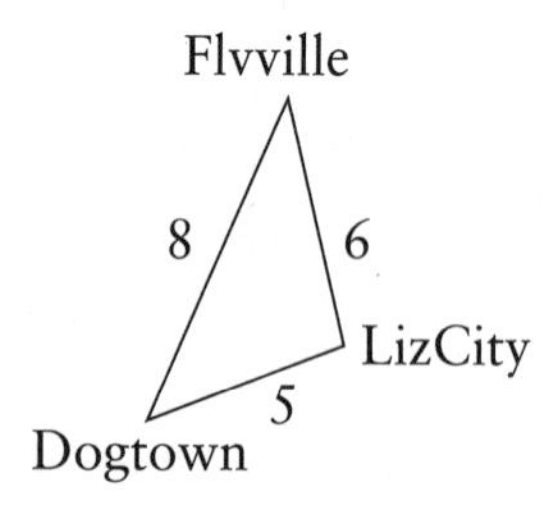

But what in a living creature supplies a valid measure of biological or evolutionary distances, assuming the role of geographic separation in our flat map? We take advantage of knowing the nucleotide sequences of the same gene from different organisms. By aligning two gene sequences and counting nucleotide differences, we measure a meaningful evolutionary distance, related to the amount of genetic change observed when the two genes (sequences) are compared. For example, here are two RNAs aligned as similar strings of nucleotides, A, G, C, and U:

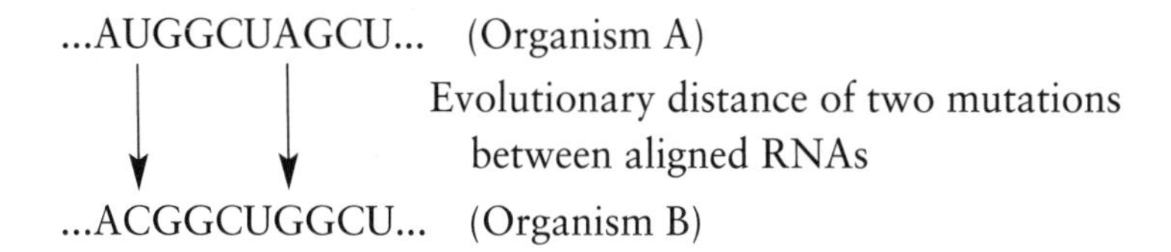

(RNA is composed of nucleotide strings, like DNA's A, C, G, and T, but its constituents are A, C, G, and U instead.) Thus we deduce that the time elapsed since the common ancestor for both these sequences has been sufficient to allow two mutational changes, or an average of one change per sequence. Now draw a map by employing these observed evolutionary distances, in which the differences have accumulated on each side during the time since the two genes were separated into different species:

→ Time *Ancestor* —< A, B

Here the distance (time) to go down the branches to sequence A or B is one unit for each side, totaling the two changes that we saw when A's nucleotide sequence was compared to an aligned copy of B's. This outcome is a simple tree, with a trunk and two (A and B) branches.

Because genes in different species are genetically isolated (with important exceptions that come up later in this chapter), the genes differ more and more after separation, as do the species that contain them. With a useful gene (which must be present in every organism), we can repeat this process to lay out a map of the distances between all forms of life on Earth. As we know the genomes of more and more organisms, we will be able to use the changes in the entire structure of their genomes to relate them. But for the moment this genomic comparison is still too difficult. It is clearer to choose one gene, or a few, and to follow their changes in sequence from one creature's DNA to that of another.

Which gene should we use, then? It is necessary to follow a gene that is not easily transferred to another creature. This condition requires explanation, because it rests on an undeniable but truly startling quality of genomes: most genomes are patchworks of genes that have accompanied the organism from earliest days, but also genes that have jumped into the genome at different later times. Any organism that we are in contact with can originate a piece of DNA that ends up being incorporated, or recombined, into our own human genome. This may seem rude and/or inopportune, but it happens by accident quite often over the long haul of evolution. In fact, there are even viral organisms that make a living by making such intrusions into our own genomes: the retroviruses. Such jumping is called lateral gene transfer, and it is startlingly frequent. One could write a useful light history of the idea of the genome by tracing the decreasing respect accorded the idea of an enduring alignment of genes. Real genomes often change randomly and particularly, frequently suffering minor invasions by others. In fact, 50% of an admirable genome (our own) is the result of this kind of recent incursion!

To stray back to the topic at hand: gene transfer between different creatures confuses the evolutionary distances we assign by counting nucleotide sequence changes. Given such jumping, the history of a creature is not necessarily the same as the history of all its genes. We want the genes used for our Big Tree to have stayed completely put over deep time, so that a gene's sequence records only the history of its present organism.

The ribosomal RNA (rRNA) gene is a frequent choice for such evolutionary studies. rRNA is wound through the functional heart of the cell's protein assembly robots, the ribosomes (see the lexicon and Chapter 17). rRNA changes slowly and is present in all cells, so that organisms can be compared across deep time. Moreover, the protein synthesis machine—of which the ribosome, with its RNA, is a complicated major part—is even larger and more complex and therefore difficult to transfer successfully. It is unlikely that grafting the front end of a Toyota Yaris onto the back end of a Rolls-Royce will yield a means of transportation as good as either a complete Yaris or a complete Rolls-Royce. Because the mechanisms in front must work with those in the rear, the inexpensively engineered Yaris is unlikely to be improved by an indiscriminate graft from the haughty Rolls. In the same way, because you can't transfer part of the huge assembly that is the protein synthesis machine, you must either take it as a whole or, more likely, not transfer it or its parts at all.

Carl Woese of the University of Illinois began making the measurements for such maps by counting nucleotide differences between rRNA genes (even before large-scale sequencing was possible) and interpreting them as evolutionary distances. The cumulative number of sequence changes increased with greater separation in time. The idea that such molecular changes roughly measure the time since two genes were one (inside an ancestor) is even older, having first been proposed by the chemist Linus Pauling and the biologist Emile Zuckerkandl, who were in fact thinking about other biomolecules, proteins. While it is tempting to count sequence changes as direct indices of time, the rate of change varies in response to factors we cannot always know. So nucleotide changes are only roughly interpretable as ticks of an evolutionary clock. The clock keeps time best over intermediate separations, because these tend to keep the history between two organisms simple.

Time's approximate, rather than exact, influence is evident in the drawing of the tree, where the present is at the periphery for all lineages. The distance to the old center is usually somewhat different along any two paths, even though the underlying times are necessarily always equal. This is because nucleotide changes are accepted into the genomes of separated creatures as arbitrary, independent events. The total numbers are somewhat different as a result of this variation. So use of the mutational distance (which is what you can actually measure) as a stand-in for time (which is a deduction and not directly observable) must be undertaken with caution. Usually, the smaller the length of the branches (depth of time and complexity of descent), the better. However, our earlier example of a distance two mutations (the distance up one branch and down the other branch to the second sequence) errs too far on the side of simplicity—it is so small that it might be inaccurate because of the variability that always accompanies small numbers of events.

But, with distances between all organisms measured by counting the sequence changes, we can ask for the diagram of evolutionary descent (the tree) that best represents the set of measured distances. The best trees must take mutational chance into account. Actually, the tree in Figure 3.1 is even a bit more sophisticated: it is the result of asking which tree would be most likely to give rise to the observed distances, given that certain kinds of mutations are more probable than others. (For example, the nucleotide bases vary in size: interchange of the small bases C and T is more likely than exchange of a small one for a large one, such as C becoming A.)

What then do we see when all this computation is finished? The answer is amazing and encouraging. Genes (or at least rRNA genes) behave just as evolution by descent from common ancestors suggests they should. These simple ideas about sequence resemblance and descent successfully order all life on Earth into a simple treelike diagram. Just by counting sequence changes, we have reproduced most of what we would have concluded by using all other information we can bring to bear about the macroscopic and microscopic look of organisms. The Big Tree is one of the (lesser known) triumphs of biological science; it summarizes and orders Earth's creatures in somewhat the way that the periodic table organizes the chemical elements.

The tree shows none of the myriad of groupings that would have seemed immediately crazy, for example, humans grouped with (having RNA sequences more similar to) butterflies to the exclusion of other insects. Among organisms big enough to see (the "crown group" at the middle right of Figure 3.1), cats group with other felines, dogs with wolves, apes with humans, and plants with other plants. Distance on the molecular tree

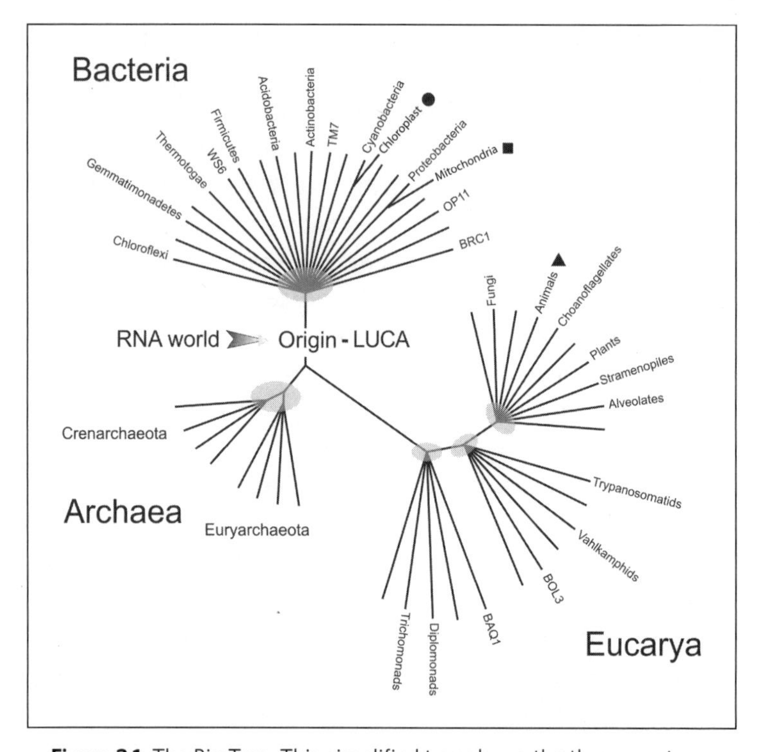

Figure 3.1. The Big Tree. This simplified tree shows the three great domains of life, defined by ribosomal RNA sequences. Grey ovoids mark areas within which evolutionary divergence is not yet confidently resolved. A square marks the mitochondrion and a circle, the chloroplast, each closely related to an existing bacterium. Humans are not resolved from other animals at this low resolution, but are within the branch indicated by a triangle. Alphanumeric names refer to creatures known only from their nucleic acid sequences. The Last Universal Common Ancestor or origin is the root of the tree, behind or before which lies the RNA world.

usually agrees with other indications of relatedness. However, molecular distances have a crucial advantage: they are available even in difficult cases, when appearance is not useful. And, finally, rRNA molecular data link all known creatures, spanning all cases in which new species appeared and still have descendants on Earth.

That descent with modification makes sense of the cousinship of most of the world's biota is a strong argument for the Darwinian process, and thereby for the descent of all of Earth's biota via genetic change from a single origin. Conversely, the tree shows that all present creatures are surprisingly related, much more so than their outward appearances suggest. The snake and the platypus and the whale have different, but related, DNA sequences that belie their outward differences.

The tree both links the various and distinguishes the similar. With all due respect, are you related to bread fungus? The Big Tree shows that you are (along with the rest of us), and tells you just how close that relationship is. The tree's distinctions among apparent similarities are notably useful for microbes. Even under a microscope, there is not much to go on in judging relationships between similar bacteria, which look like similar spheres or oblongs. However, molecular sequence data resolves them easily. Two bacteria, for example, a Bacillus from the soil and a human pathogenic Clostridium, are about as different as a human and bread fungus. In fact, most of the length of the tree's branches is in lines leading to microbes. Therefore, most of Earth's diversity resides in microbes. That is, most of evolution has led to microorganisms, not to the more familiar world of animals and plants and insects. It's a microbial world—now as at life's beginning.

Furthermore, the reach and bushiness of the microbial branches of the Big Tree reflect the fact that the microscopics really are more diverse than we macroscopics. We make our living on the Earth's surface by eating other creatures, or, if we happen to be plants, by fixing carbon dioxide. The microbes instead live in a dizzying variety of environments on the surface, inside rocks, in boiling springs and oceanic vents, and deep inside the Earth. They metabolize methane or sulfur or metals or nitrate or hydrogen. In other words, they express the variety of their genomes, seen from one point of view in the tree of their rRNA sequences, but also in their varied styles of living.

The tree's large-scale structure suggests that all earthly organisms are members of three great domains: the Archaea, the Bacteria, and the Eucarya. A domain is a set of creatures joined by shorter branches of the Big Tree and who thus fundamentally resemble each other (as indicated in Figure 3.1). These three domains, or lobes of the tree, themselves all trace back to a common ancestor of all surviving earthly life, sometimes called the Last Universal Common Ancestor, abbreviated (and pronounced) LUCA. Two domains are entirely microbial: the Bacteria and the similar-appearing (but molecularly distinct) Archaea. The third domain, in which we reside, is the Eucarya (cells with true nuclei), which includes protozoa, fungi, insects, and most everything you can see in a zoo or a museum. Near the central root of the tree is the LUCA, whose offspring radiated out into all we now find living on the planet.

The radiation of the three domains from common ancestry is actually evident from independent resemblances between organisms. For example, there are 64 codons in the genetic code. (A codon is a sequence of three nucleotides that code for an amino acid; see Chapter 18.) By one possible reckoning, this allows for 1.5×10^{84} ways to assign nucleotide triplet codons to amino acids and the stop and start signals for proteins—that is, 1.5×10^{84} genetic codes. This is more genetic codes than there are fundamental particles in the known universe. Thus, the use of only an inconceivably infinitesimal fraction of these possible codes, similar in all earthly organisms, points strongly to the acquisition of one particular code from one common ancestor.

One of the most impressive results of the Big Tree is that two domains of life are today living together. To see this we must look hard at microbial details from the tree and compare them with rRNA sequences from us lumbering, multicellular types (denoted by the triangle in Figure 3.1). Now consider symbiosis, the invasion of one organism by another, as a source of novel creatures in evolution. This startling notion was devised by a Russian lichenologist, Konstantin Merezhkovsky, around 1905, when it was realized that the organisms called "lichens" were a fungus and an algae living together. The same theme occurred to Ivan Wallin, a University of Colorado microbiologist who drew considerable abuse in the 1920s for persistently claiming that the mitochondria (energy-providing compartments) in cells like ours are actually potentially free-living bacteria. While these ideas were incorrect in their details, it is now clear that the metabolic capabilities of Eucarya were in fact the result of invasion by a microbe that took up residence and became our mitochondrion. Similarly, the chloroplasts that make plants green, capture light, and fix atmospheric carbon are the durable, useful remnants of an ancient invasion by a photosynthetic green microbe. These notions of invasion and symbiosis were in recent times revived and promoted by the biologist Lynn Margulis of the University of Massachusetts.

Looking at the Big Tree, we can see symbiosis, startling though it may be, borne out in a convincing set of sequences. The rRNA within the mitochondrion is closely related to that of a type of bacterium (denoted by the box in Figure 3.1). The rRNA in plant chloroplasts has a sequence related to that of the rRNA of a photosynthetic bacterium (denoted by the circle). These rRNA sequences testify unmistakably that an idea almost too far-fetched for a horror movie—the fusion of two organisms to make a more capable one—is not only true in principle but also true inside each of us. Both animals and plants are evolutionary condominia—ancient collaborations still alive in every eucaryotic cell, like the plant cell in Figure 3.2.

Or alternatively, to illustrate how significant is the resolution between organisms seen in the tree, reflect on the following paranoid rhapsody. There are, in secret among us, disguised inhabitants of another world. They are able to take familiar forms, appearing just like everyday earthly creatures when

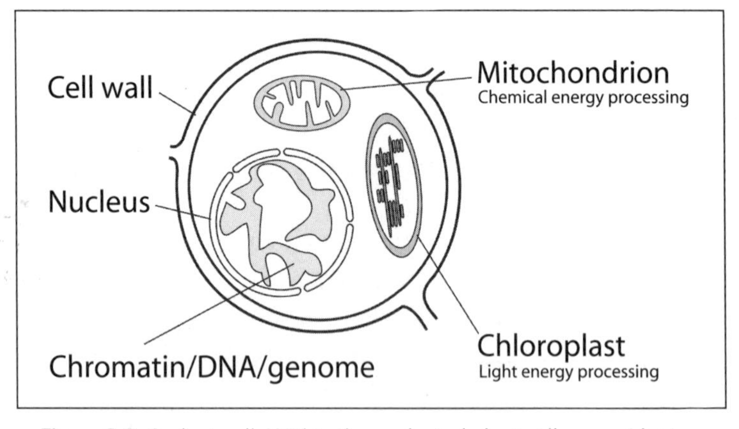

Figure 3.2. A plant cell. Within the nucleated plant cell are resident subcompartments, the mitochondrion and the chloroplast, that originated as other organisms.

A paranoid postscript

Or are the aliens sufficiently cunning to prevent their ribosomal RNA sequences from ever entering sequence databases? Or perhaps they have carefully provided themselves with sham ribosomal RNA genes that match their treacherous exteriors, so they do not attract the wrong kind of attention. Uh-oh.

that is expedient for them. But underneath they act from minds bearing an unalterably alien vision, bent on the clandestine occupation of our planet and the displacement or annihilation of the creatures that are truly at home here. They await only a signal flashing from the interstellar darkness to reveal their true, malevolent natures.

Fortunately, there is strong evidence against not only this but any other description of the Earth's creatures as progeny of more than one origin, creation, or world. This is, of course, the Big Tree of rRNAs itself—its impressive wholeness shows that written deep within our molecules is evidence that we are successive elaborations of a single kind of ancestor. There are no numerous malign, silent outlanders.

All the world's rRNAs align because they all stem from a single source. This view immediately places a number of constraints on our origins. We are not created from the same bin of identical parts. Nucleotide sequences from different organisms, even if taken from the same genes, differ and can differ greatly. It is equally clear that we were not created separately, like automobiles that, despite similar parts lists, were designed independently to meet only partly overlapping goals. Each earthly organism does not represent a new solution to the problem of being alive. Instead, our parts have unexpected bumps and hollows that are the same in all for no functional reason, and that therefore testify to our interconnected paths that diverge from an ancestor, embodied in the tree.

Now we can turn toward the RNA world. Life's origins necessarily trace to the simplest organisms because they are the ones most likely to originate from nonliving material. Therefore people often look to bacteria, as the simplest current life forms, for a model of the first organisms. But today's bacteria have 4 billion years of evolutionary theater behind them and need not resemble the first replicators that shifted into high gear by inventing Darwinian evolution. (See the following chapters, particularly Chapter 5.) However, the underlying point must be accurate: the earliest organisms resembled simple chemicals more than any later living thing did. Thus the Earth has been dominated by the simplest organisms from the start, and the Big Tree reaffirms an old and persuasive deduction. If we ever find another planet with varied life, it will be similarly treed.

This is not necessarily to say that there have not been even more startling kinds of creatures walking, creeping, and sloshing on the Earth. It is a peculiarity of evolution (see Figure 3.1) that over a sufficiently long view, one branch of living things wins and the others go extinct. As a matter of fact, it is estimated that 999 of every 1,000 species that have ever existed on Earth are now extinct. As we look around and count those who are similar to us (who are all we can see still standing), it is well to bear in mind that there are countless other beings who might have appeared incomparably weird, had they survived. Over deep time, history is written by one group of related winners. And the losers might go back to ancestors with surprising differences from us, the related inheritors of the Earth today.

Extinct humans

Extinct types notably exist even for humans. Many other types of human beings probably shared the Earth at one point, perhaps grumbling that "they" did not show proper respect or had no decent music or proper morality. We know this because humans are relatively large and bony, and so leave remains that last. The bones of the Neanderthals yield DNA whose sequence reveals that they were deeply different from and only distantly related to us. We will shortly know much more about this difference: there is a Neanderthal genome project that intends to recapture their entire DNA genome to compare it with modern *Homo sapiens* DNA. Up to recent times, about 35,000 years ago, the Neanderthals and our immediate human ancestors lived side by side in Europe. Think of it—a visibly different kind of human with a distinct culture (as evidenced by particular burial customs and distinctive tools). One would like to have been present when these neighbors met, say, out hunting. But we won and they lost, and we now write the histories, calling these strongly built humans—whose brains were the size of ours, or even a little larger—"primitive" and pointing to their distinctive brows and chins. There is no avoiding the fact: the present holds only a very partial sample of those who have lived and perhaps thought of the world as rightfully their own. Even Neanderthals, intelligent beings so like us that we must strain to define our differences, are gone forever.

In Figure 3.1 we are looking at the Big Tree from "above," with the most recent branches closest to a circumferential line drawn around the entire tree. So to find the region most relevant to a discussion of ancient ancestors, we must look through the bushy parts to the innermost parts of the tree. Here we make an unexpected observation: a considerable amount of

The position of the LUCA

The position of the RNA world on the tree is quite certain: for example, it is very likely that the LUCA had ribosomes and was capable of translational synthesis of proteins. This is because every cell in all domains of life has recognizably related ribosomes today. The only imaginable explanation for this fact is that the Last Universal Common Ancestor bequeathed its ribosomes to all modern organisms. Thus the RNA world, before the rise of the proteins, lies behind or before the root of the tree—before the time of the LUCA.

change occurred in rRNA genes before modern species (branches) began to be observed. Why did the LUCA (or its rRNA genes) leave no modern descendants immediately? Carl Woese has interpreted this blank zone as a time before modern information processing existed, so that stable species (which must pass their genomic information accurately to descendants) could not be formed.

In any case, the common ancestor of all life on Earth is near the center of the tree, where the lines come together. All known rRNAs are obviously related (that's how we align them to make a tree), so they must join at the center. But in the mysterious empty region at the center of life's divergence, ribosomes were changing a great deal without leaving descendants we can recognize today. The first branches on the three limbs all lead to modern cells, which are today's DNA microorganisms. Somewhere inside the first branches on the three limbs of the tree, defined by changes in RNA molecules, is the passage to the RNA world, where we are headed.

As you can see, the first organisms around our destination are all microbial, so we are almost certainly also looking for a microbe, the RNA cell or ribocyte. Because all modern organisms have DNA, the LUCA probably was a DNA organism. Therefore, we are headed downward behind and before the LUCA, to an RNA world that flourished and died before the Big Tree began branching.

Readings

"The deep roots of eukaryotes." S. L. Baldauf. *Science* 300: 1703–1706 (2003).
The tree is still being filled out and resolved by new discoveries; in particular, what has happened since single eucaryotic cells began to join into organisms is becoming clearer.

"A molecular view of microbial diversity and the biosphere." Norman R. Pace. *Science* 276: 734–740 (1997).
Details of the Big Tree, with emphasis on the way molecular data have changed and are changing our view of microbes.

4

A Dance of Atoms

> Writing about music is like dancing about architecture.
>
> —Elvis Costello

Evolution appears to be a simple idea—perhaps having hugely diverse, continually branching consequences, but behind all the complexity, an elegantly simple idea. One facet of its luster is that it is even simpler than it first appears. Evolution's necessities are so closely linked to basic tendencies of matter that, given a few minimally demanding requirements, it becomes well nigh inevitable. Pressing Elvis Costello's simile in a different direction than he really intended, writing about evolution *is* like dancing about atoms. Evolution's link to the underlying properties of matter is the subject of this chapter.

Start by imagining mountains. Mountains are compelling because they capture the "otherness" of our world with a universal urgency. Forces beyond our ken, times beyond our lingering, shaped these impressive towers. Perhaps such forces shaped everything. We cannot help but feel that we are far from understanding such leviathans, so we are suitably inspired and exhilarated. In the idea of mountains lies a profound truth that we need, about life's means. Mountains capture a frequent way of thinking about chemical reactions.

Let us now, in our imaginations, take the mountains inside our bodies, as a metaphor for the durable chemical ticking inside each of us and in our genomes.

You are standing in a high meadow in the Rockies or the Alps. Across the way, snowy peaks rise, streaked by steep gulches, paved with glinting, icy surfaces. The skyline is jagged, rapidly declining from rocky peaks to passes. At these low points, meadows slope up to the sky, and a committed person on foot might cross to the invisible far side. Crossing these passes provides the image we need.

In this mountain landscape, we are imagining something like the surface that directs a chemical reaction, called a free energy surface. The low passes, the easy crossings, correspond to the probable chemical reactions. Were atoms and molecules to elect these accessible routes, they would likely react, that is, cross to new chemical arrangements in the next valley. The higher passes offer less probable passage (less probable chemical reactions), because reaching them requires activation energy for climbing, and so they are more rarely traveled. Peaks are places you seldom go during normal travel, and only with good and resolute reason. Seen with a chemical turn of mind, different passes and crossings might be attacks by an electron-hungry reagent on different crowds of electrons (negative charges) within a reacting molecule.

Everyone who has performed chemical reactions—with a home chemistry set or in the sanctioning name of education within a university laboratory—instinctively understands one important property of this picture. You construct your reaction according to directions, and on good days you get mostly the expected product. But in a reaction of any complexity the target product is always accompanied by alternative products that make it impure to some extent. (There are exits via other mountain passes than the intended one.) If you do something inept or mistaken that changes the starting point, you will find yourself taking an unanticipated route, arriving at an unrecognizable mixture in which the desired product is dilute or unrecognizable.

Chemical reactions can be optimized to minimize side reactions, as they are, for example, in chemical manufacturing plants. But as soon as your reactants are particles more complicated than single ions or atoms, there are always side products, and often in very substantial amounts. If the intended product of a chemical reaction is the "right" outcome, there are usually "wrong" outcomes as well. More complex reactions and more complex reacting chemicals, all else being equal, always have more possible side reactions and more wrong outcomes. To think otherwise is to imagine a free energy mountain range in which peaks all around soar into the clouds and out of sight, leaving only one low pass in view in any direction—a very, very rare landscape indeed.

Therefore, to do chemistry with molecules is to do it incorrectly to some large or small extent. If this sounds heretical, it may be because this essential point is usually concealed from students by a conspiracy that issues mute, magisterial recipes in which only simple outcomes are mentioned. These are called lab notebooks. However, aberrant chemistry is of particular importance to living creatures, who must do exceedingly complex things (requiring many chemical reactions) and get them overwhelmingly right. A level of precision that would earn you a grade of A+++ in a synthetic chemistry course would be an exponentially reverberating disaster for an organism replicating a complex genome.

To see why, consider what it takes to reproduce. Without significant error, we humans must replicate our own genome—

the recipe for a human, a six-gigabyte nucleotide text—and get it overwhelmingly correct. Such mass production is by now a routine task with silicon, plastic, and steel, but rather harder using meat. Suppose you are a chemist of surpassing skill, who can perform reactions yielding the expected product 99.99% of the time. (This is really exceptional accuracy, but I know you, and you can do it.) About once in 10,000 reactions, in other words, you make an error. In making a human genome, you would make about 300,000 errors. It may help to imagine the display in a sports stadium, a Jumbotron a hundredfold bigger than the panel on your laptop, but with the same resolution. Now imagine that 300,000 of its pixels have some odd, clashing color. The kind of (im)precision you would see is something like the basal error level of nucleic acid replication in today's simplest enzymes.

As a result, even with our highly evolved chemical error-correcting precautions, there is about one serious (potentially lethal) error (mutation) per genome replication. This will almost never bother us, however, because we have at least two copies of the genome. We can frequently use one (e.g., the old, less error-ridden one) selectively as a reference and navigate around a single isolated error. To be complex creatures inescapably means that we have found similar exceptionally good ways of working around the inevitable "chemical" errors. We, in our role as replicators, nearly always take the correct pass out of the valley, or at least switch over to the desired trail if we start off in the wrong direction.

But a primitive creature is very prone to error and cannot avoid it with sophisticated correcting devices like those we use. As an evolutionarily relevant illustration of the fundamental origins of error, recall the standard DNA nucleotides A, T, G, and C. The GC base pair (guanine-cytosine) is the

Watson-Crick norm in DNA, but the GT base "mispair" (guanine-thymine) is only slightly less stable and will necessarily occur a significant fraction of the time as a normal DNA nucleic acid sequence is replicated. Thus the nucleobase G "should" pair with C during replication, but a small fraction of the time it will pair with the abnormal partner T, a potential mistake in any genetic text. Such errors can be corrected, but a more primitive replicating system may not be able to do so accurately.

Let's summarize. Matter that has organized itself to replicate will have a much, much better chance to persist by making more of itself. If it replicates, it will necessarily replicate with errors, producing descent with variation. The ingredients for Darwinian evolution will be present automatically, given that there is a competition or selection. Competition arises almost universally as a contest of speed of reproduction, of efficiency in finding the ingredients for replication, or indeed of accuracy of replication itself. Darwinian evolution, according to this summary, arises from the logic of reproduction, and from the elementary properties of chemical reactions. These statements become steadily more compelling as complexity increases, and the requirements for evolution will almost certainly be present in any living system sufficiently complex to replicate. That is why a diverse earthly menagerie, including us and ours, is here to be analyzed.

Readings

The Blind Watchmaker. Richard Dawkins. W. W. Norton, New York (1986).

This volume, cast in a thoroughly reductionist frame of reference, is the most compelling general-purpose book on

Darwinian evolution written during the last decades of the twentieth century, unless of course . . .

The Selfish Gene. Richard Dawkins. Oxford University Press, New York (1976).
. . . this *is the best general-purpose book on evolution. The second edition (published in 1990) has worthwhile additions.*

5

Allegro Agitato: The Origin of Life

> You can lead a dead horse to water, but you cannot make him drink.
>
> —Alan Austin Lamport, mayor of Toronto

This book is not primarily about the origin of life, but the origin is an essential gateway to the RNA world. However, life's origin is necessarily the most distant and least known step in the rational description of earthly life. Put another way, the conceptual gap between geochemicals and things alive seems the most formidable leap required in the telling of this story. Should nonrational explanation ever play a role, the smart money would bet (as it often has) on its taking a role in origins.

Nevertheless, this is not for the most part an account of life's origins, because the RNA world cannot occur at the origin, as a property of the first living things. RNA is a complex, fragile molecule (see Chapter 10) that can only exist after descent from a world in which there has been considerable prior evolutionary progress. In particular, RNA is relatively unstable and requires a metabolism that creates it, replicates it, and possibly repairs it in case of chemical difficulties. Thus the RNA world and its biological inventions can only be plausibly envisioned as a productive but relatively late era, arising some time after the very first living, evolving things appeared. The RNA world is surely the apex of a first golden age, built on an already-extensive molecular culture.

Life on Earth has come to the present through deep time; it has persisted here for some billions of years, just under a third of the history of the universe itself. To make sense of life's history, as a first item on our agenda we need a way of grasping those vast, chilly eons. Perhaps "grasping" is too strong a word; but we at least need a way of appreciating potentially mind-boggling spans of time.

In order to put such time in terms of personal experience, imagine a hundred years—a possible human lifetime, which seems comprehensible even if we have not closely observed anyone through such an exceptionally long life. Accordingly, the gap between us and the hundred-year-ago world just after the year 1900 is the smallest we will consider. But who today knew Wilhelm II, the emperor of Germany and an inescapable Western political figure of the early twentieth century? We need longer times and better-known symbols to span them. Usually larger things move more slowly, so in trying to visualize longer periods we should generally favor bodies larger than persons or nations.

How about mountains? Mountains come and go on time scales of tens or hundreds of millions of years, as continents move and collide, following the logic of plate tectonics. So even the lifetimes of mountains, or ranges of mountains, though they seem vast and permanent, are still not enough to easily measure out the history of Earth's biota.

If continents and mountains are too small and transient for our purposes, let's use something yet larger for our measure. Only the ancestors of today's landmasses, the supercontinents, are larger distinct features of the Earth. But the well-known supercontinents Pangaea, Gondwana, and Laurasia are still too young. We must go back one billion years ago (1 Gya, a gigayear ago, 10^9 or 1,000,000,000 years ago) to Rodinia, the first supercontinent and the mother of all surviving continents. The time required to break up Rodinia, reform the supercontinental Earth, form and break up Pangaea, form and break up Gondwana, and watch the continents sail to their current sizes and positions—that is 1 Gya and a unit commensurate to the history of life on Earth.

Judging from the satellite-borne Wilkinson Microwave Anisotropy Probe's data on the leftover radiation from the Big Bang, the bang occurred 13.7 Gya, with an uncertainty of only 200 million years. A few hundred million years later, the first stars began to shine, and the Milky Way galaxy, our home, came into existence. In fact, the apparently oldest stars in our galaxy have an age indistinguishable from that of the universe itself, so the oldest parts of our galaxy were formed fairly early (though this time estimate has much greater uncertainty). The figure of 13.7 Gya is an important number because it supplies an upper limit on the age of *everything*. Nothing can be older than the First Singularity of the universe. Thus when it is suggested seriously that life on Earth has extraterrestrial origins (and it *is* suggested seriously), no matter what distant parts of the universe life might have come from, it cannot be more than about threefold older than the Earth itself. This is an important fact—whether or not you are entertained by the idea that we are extraterrestrials.

We can be more precise about the age of the solar system, including Earth. By measuring the decay of uranium isotopes to lead in meteorites we know that the rocky material of our solar system congealed at 4.55 Gya (with an uncertainty of only about 1%) in the wake of local supernova explosions that created such heavy atoms as oxygen, carbon, and nitrogen. A period of intense bombardment followed, during which any oceans would have been boiled and vaporized by the furious impacts of extraterrestrial objects on the early Earth. The simultaneous melting and explosive redistribution of the crust would have made the land similarly unstable and inhospitable. By dating impact craters on the Moon, which was buffeted by the same process, we know that things had quieted down (at least relatively) by roughly 4 Gya. For the first time, a planet with stable oceans and extensive crust was available. This time occurs between two of the modest dots on the timeline of Figure 5.1.

What about life? The empirical evidence for the origin of life is in flux. Earlier dates were supported by isotope fractionation data and fossils that are now disputed; what was once interpreted as an early sign of life may in fact be due to simpler chemistry and physics. It now seems that the most ancient, minimally controversial sign of life on Earth is deposits of

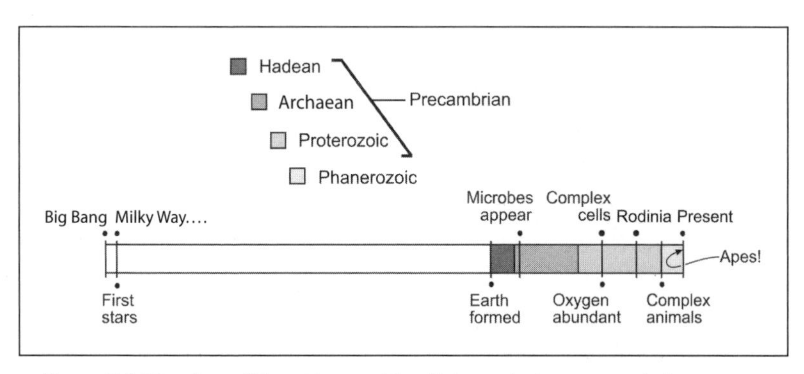

Figure 5.1. Timeline of the universe. The distance between events is proportional to time.

carbon atoms. Carbon has heavier (^{13}C) and lighter (^{12}C) stable isotopic atoms, and this old carbon, dated to 3.85 Gya, has the carbon-isotope ratios (enriched in ^{12}C) characteristic of biochemistry. Even more vivid evidence is provided by stromatolites, layered hassock-like structures built by colonies of microorganisms in shallow water and fossilized to display a characteristic layered rock signature in their cross sections. These can be dated by isotope measurements to 3.45 Gya. However, because these stromatolites were built by relatively advanced organisms it seems likely that life, at least bacterial life, had appeared toward the middle of the first billion years of Earth history. This chronology is consistent with apparent isotope fossils at 3.85 Gya and implies that organisms were abundant a maximum of a few hundred million years after the primal bombardment of Earth slackened.

As we saw in Chapter 3, there are three broad styles of life on Earth—organisms more similar to each other than they are to other styles of organisms: bacteria, archaea, and eucaryotes. Bacteria and archaea are mostly single-celled and have no nuclei; the DNA genome swims, unrestricted by a membrane boundary, in the cell's interior. Eucarya have nuclei and other subcellular membrane-bounded compartments; they are the building materials for almost all complex, multicellular beings. Nevertheless, all three styles of creature decorate the same tree (descend from a common ancestor) and all may be traced back similarly close to the beginnings of life.

Stromatolite formers were simple, presumably bacterial creatures. Eucaryotic (nucleated) cells apparently as complex as our own do not appear obvious until 1.6–2.1 Gya. Thus life was simple and microbial, single-celled, through billions of years of history, waiting for the invention of cells capable of a more complex style of life. This complexity appeared about

2 Gya, more or less simultaneously with the rise of oxygen in the Earth's atmosphere and the appearance of an efficient oxidative metabolism. Simple nucleated and nonnucleated cells were the citizens of Rodinia. Life in the form of our single-celled ancestors rode the diverging continents and was undoubtedly shaped in yet-unknown ways by these long, ancient crustal wanderings.

The modern eucaryotic cell arose through invasion by expert bacterial metabolizers that became the cell's mitochondria, and for plants, through invasion by blue-green cyanobacteria that evolved into the oxygen-generating chloroplasts. Microbes often ingest smaller creatures that they intend to digest. In the descent of eucaryotic creatures like us, the intended prey very occasionally took up residence instead of becoming lunch. The mitochondrion and chloroplast can, as a result, be placed on the Big Tree alongside ordinary easily recognizable bacteria and blue-green cyanobacteria. (See Chapter 3 for more of the story.)

Despite this dramatic eucaryotic elaboration, it is more than a billion years later, at 0.6 Gya, that the earliest multicellular animals can be found: the simple saucerlike creatures called Ediacara, named after the Ediacara Hills in Australia. About 0.54 Gya modern complex multicellular animals left fossils (in the Burgess Shale of the Canadian Rockies) that distantly resemble modern multicellular forms.

From a human point of view, only about 500 million years now intervene before another crucial innovation. Only 0.006 Gya (6 million years ago), having come out of the sea and diversified, a small band of exceptional African apes began dangerous experiments with language and technology. About 0.00001 Gya, descendants of the upwardly mobile, now relatively hairless ape invented agriculture and took up a settled lifestyle in the Middle East. The farmers' closest surviving relatives became modern chimpanzees. In the timeline in Figure 5.1, every kind of thoughtful ape that resembles us has lived and died within the width of the rightmost boundary line.

This brief Earth history teaches us that simple life arises easily (because it did so relatively quickly). In contrast, animals with complex body plans take much longer (because they took gigayears to appear, until the Ediacara). Intelligent life, necessarily complex (because it requires a brain or some other intricate organ of analysis), follows a few hundred million years after the rise of varied body plans, and therefore probably requires newer and simpler developmental tactics than did complex bodies. So, somewhat surprisingly, once you can make a complex body, with cells devoted to different functions, a nervous system follows relatively easily. Most likely, understanding animal intelligence will follow with ease once we understand animal development. Nevertheless, high intelligence of an abstract kind deserves respect—it is extraordinary and has occurred among the descendants of only one of the tens of millions of speciations that make up the Big Tree, populating the contemporary Earth. Clearly we will expect a bit of unusual developmental magic along the evolutionary line leading to us and ours. Yet we should maintain some perspective; the Earth may have seen more intelligent creatures than us—creatures who, through lack of food or inadequate lust for reproduction, are now extinct.

Let us now return to that time between the slackening of bombardment by fragments of developing worlds (4 Gya) and the distinctive appearance of biogenic relics on Earth (3.85 Gya). This is a world governed by the rules of chemistry. It has sources of energy and biomolecules, but these will run down according to the second law of thermodynamics as the history of the world ticks on. The second law says that closed systems in which real events occur must become disordered: entropy must increase. What can persist in such a declining world, where all is cooling and spontaneous, undirected chemical consumption?

Consider something that consumes its surroundings and converts them into more of itself—something that replicates. Such a self-focused being can, uniquely, survive and even increase in numbers and extent, even as all around it declines. In fact, even as it contributes to decline by consuming its more inert surroundings, it defies that pervasive decline by itself flourishing. Replicators have just those properties that allow them to thrive when all other things are necessarily progressively lost. Therefore, life will be a salient quality of a world wherever and whenever matter can take a form that replicates itself. Replicators possess a unique, soft durability that enables them to outlast the very stones.

Because we are here, we know without doubt that matter potentially takes a form that can replicate itself. This makes the emergence of life, descent with variation, and in fact Darwinian evolution all plausible if not inevitable. The sparse chemical details of the first appearance of living and replicating things deserve their own separate discussion, and that follows in the next chapter.

Life, with its ability to consume precursors and make more of itself, thereby flouts the date with the undertaker implicit in the second law of thermodynamics. Life ingests its surroundings—perhaps very selectively—and uses the energy thus obtained to avoid decline into disorder, instead making more of its own information-rich, highly ordered kind. Of course the second law—being, after all, a Law—is not violated. Replicators export more decline than they take in, so that

What is life ?

A further note on the nature or definition of life put forth in this chapter: this point of view suggests that "following the water" or "following the energy" or other NASA-approved ways of stalking new life may be of immediate use, but such an approach could prove blinkered in the end. The essence of life is descent with modification. And furthermore, this quality is self-selecting in that it allows life to survive the general decline of its vicinity by making more of itself. Thinking in this way, replication is life's ultimate chemical and physical survival strategy. There may be life, therefore, without water or any other familiar chemical sign. When we follow water, we will automatically be led to watery life, and we will be tempted to read this as confirmation that all life is like us. But this can easily be an error. This error potentially becomes more onerous as we move to simpler living things, who need only meet the most basic requirements, and as we move to alien environments, where our chemical intuitions will be mistaken.

overall, the universe declines as mandated by the law. But life finds a bright local exception to general chemical decline in the replicator and its descendant genetic texts.

Replication, precisely that property that is required for survival, is also required for evolution. The inevitable mutations (see Chapter 5) allow variants to occur. These are selected, and we are off and running, with the fastest or most robust running farthest. Thus if matter replicates, evolution appears, and the replicators not only persist but also evolve into varied forms.

There is a "definition of life" implicit here, and it is one generally favored by molecular biologists like me. Of course there continues to be a lot of zealous and inconclusive discussion of such definitions. In the spirit of ecumenical reconciliation the foregoing account embraces an enthusiasm for replication (and variation and Darwinian selection, as its automatic consequences), metabolism, and individuation (cellularization), integrating them into one account that considers them all to be essential.

Reading

"The origin and early evolution of life: Prebiotic chemistry, the pre-RNA world, and time." Antonio Lazcano and Stanley L. Miller. *Cell* 85: 793–798 (1996).
A broad but brief review of work on origins, with many useful references, told in Stanley Miller's inimitable voice, now stilled.

6

The Winds That Blow through the Starry Ways

No scientific subject holds more surprises for us than Biology. Foremost is the surprise that life exists at all.

—Thomas Gold, *The Deep Hot Biosphere* (1999)

In the 1899 poem "To His Heart, Bidding It Have No Fear" by W. B. Yeats, the starry winds are a subject of dread. However, in this chapter the same winds blow in the opposite direction; they are reassurance and therapy for physicist Thomas Gold's surprise, a century after Yeats, at the existence of life. The subject of this chapter is the chemical origin of life's materials. This discussion has been separated from that of the previous chapter because when we speak of life's "origins" we may mean two very different things. On the one hand, the origins of earthly life can mean the expansion to today's diversity from simpler beginnings. This is a comparatively easy question to address, because it rests mostly on descent with modification and selection—a powerful recipe for finding effective solutions to biological problems, even to problems that are unimaginably complicated. (See the numerical example in the next chapter.)

The other meaning of "origins" poses a more abstruse problem, relating as it does to the appearance of the first living things from nonliving origins, the process sometimes called abiogenesis. This question is more difficult to address because there is a different kind of functional gap that life must span. Moreover, because these events happened billions of years ago, evidence bearing on them is hard to find.

In fact, this process is also called spontaneous generation by those viewing it from a modern perspective, and spontaneous generation was ruled out, over short times and under ordinary modern conditions, by the famous experiments of Louis Pasteur. In 1862 Pasteur showed that no life arose even in a nutritious broth if the vessel containing it was sufficiently well sealed that preexisting organisms could not get in. The broth could even be open to the air; but if an organism had to navigate a fine, tortuous tube to get into the nutritious stuff, the result would be the same. Organisms therefore are the improbable outcomes of ordinary chemistry acting over time spans and in contexts that we frequently encounter. So I have set this chapter by itself, to highlight both the problem's unique importance and its difficulty. Furthermore, seeking the sources of living chemistry will lead us to answers that are intriguing in themselves.

How did life on Earth first arise? There is a library full of discussions about this question, and you might want to read from it. (Start with the reading list at the end of this chapter.) However, in one foreshortened dimension, the topic looks simple to molecular geneticists. As we saw in the previous chapter, the essence of life is the ability to transmit the information needed to build creatures like you (your genome) to descendants. Without further constraints, you are thereby undergoing Darwinian evolution—descent with mutational chemical errors and selection. Accordingly, your kind rises above the decay implicit in ordinary chemical processes—for example, you become more complex if your environment requires it. The problem of the origin of life, interpreted broadly, becomes that of the origin of genes and inheritance. So what was the first system that could replicate, and thereby automatically evolve?

The answer, as is so often the case, was first approximated by Charles Darwin. In a letter to Benjamin Hooker 12 years after the 1859 publication of *On the Origin of Species,* he wrote: "But if (and oh! What a big if!) we could conceive in some warm little pond, with all sorts of ammonia and phosphoric salts, light, heat, electricity, &c., present, that a proteine compound was chemically formed ready to undergo still more complex changes, at the present day such matter would be instantly devoured or absorbed, which would not have been the case before living creatures were formed."

Of what and where might the first replicators have "chemically formed"? A rough guess that still guides experimentation came from Alexander Oparin in the Soviet Union and J. B. S. Haldane in England, beginning in the 1920s. The idea was that chemicals would accumulate, perhaps in the ocean or at its margins, to produce a soup sufficiently rich to spawn the first living forms. We now know that the infall of material from space would also have brought many organic compounds to the Earth's surface and that some of these could have survived impact. Haldane's "soup" long ago entered the lexicon, and cans labeled "primordial soup" are a popular image illustrating studies of the origin of life.

But note well: The primordial soup is not to everyone's taste, with some workers claiming that such a liquid is pure fiction. In their alternative scheme, ancient organisms themselves made every molecule they needed. This alternative, "autotrophic," explanation for the origin of life can be followed through the reading list or in the writings of Günter Wächtershäuser or Christian de Duve. This variation would greatly expand the scope of my message, but in fact change it only slightly, so I will stick to the soup menu. Heterotrophs that eat soup have the logical advantage of being simpler, because they take more order (in the form of more sophisticated chemicals) from the environment. They therefore more easily span the gap to things that are nonliving. Because this abiogenic gap is already so wide, in this account I cast our lot with early heterotrophs.

So what is the souplike situation in which a natural collection of chemicals could have spawned a simple replicator? We have only fragments of a chemical route, and those fragments we have favor the notion that some of our essential genetic chemicals are stable and easily formed from abundant natural materials, using well-known, probable chemistry. This makes it highly likely that we and other earthly living things were, at least in part, cooked up from natural ingredients. However, chasms still gape between these suggestive yet isolated observations, and much investigation still remains to be done. A full recipe for the vital soup has yet to be written, despite hopeful early tastes of it.

Most famous among these tastes is the experiment published by Stanley Miller, a founder of the field of prebiotic chemistry, when he was a graduate student in Harold Urey's laboratory at the University of Chicago in 1953. Urey thought that the primitive atmosphere would have resembled the composition of the gas giant planets. So Miller boiled water and channeled the vapor to a flask containing methane, ammonia, and hydrogen. To provide energy, this synthetic primitive atmosphere was subjected to a continuous high-voltage spark discharge—lightning in a bottle. Remarkably, within a few days, the flask was coated with a yellow-brown residue—evidence that abundant synthesis of more complex carbon compounds had occurred. These included amino acids, the monomers needed to chain together proteins. The simplest (and most abundant) of these, glycine, was the product for more than 2% of the methane initially used in the experiment.

All told, some 13 or 14 of the 20 amino acids found in biological proteins (as well as many other compounds) have been obtained from this experiment and similar ones. Thus the constituents of proteins are a major product of undirected chemistry. To be fair, hydrogen or methane (reducing conditions, in the jargon of the field) must be present in excess to yield these results, and there is no certainty that this was the case on the early Earth. However, we really don't need to settle this particular question for Earth: the universe as a whole favors reducing conditions—it is filled with its favorite and simplest molecule, hydrogen, a ubiquitous constituent, abundant everywhere in space.

The Miller-Urey results are the outcome of well-known chemical processes; aldehydes (like formaldehyde; Figure 6.1) and nitriles (like cyanide) are created by spark-driven rearrangement of the starting materials. These react to yield amino acids via a named route: the Strecker synthesis. I mention these chemical details because we will need them later to draw the essential conclusion about favored chemical routes.

Another encouraging sign comes from Joan Oró (and A. P. Kimball), whose results were published in 1961. Five molecules of HCN (hydrogen cyanide, a deadly poison to large aerobic creatures like us) at high concentrations polymerize in a bit of ammonia to yield insoluble black gunk, but also a large amount of free adenine (A, one of the standard bases A, G, C, and U/T) of DNA and RNA. Even more adenine can be released by taking apart the gunky products with acid. So an essential building block for a nucleic acid genome is the result of easy, spontaneous chemistry. And the required starting material, hydrogen cyanide, should be easy to come by on an early Earth, as we have just noted.

Even better, slight modifications of this scheme will yield the other bases. On the other hand, the efficient manufacture of adenine requires truly high cyanide concentrations, even beyond those that might be imagined in the wake of ancient lightning. In the laboratory, this need can be met by freezing solutions of HCN, so that the last bit to freeze contains most of the cyanide that began in a larger volume. Here is a theme we'll pick up later in the chapter: though a plausible reaction exists, a special environment is required to make purine bases using the easy route discovered by Oró.

This account of origins notably requires that we come from and through an environment that would now kill us. The early

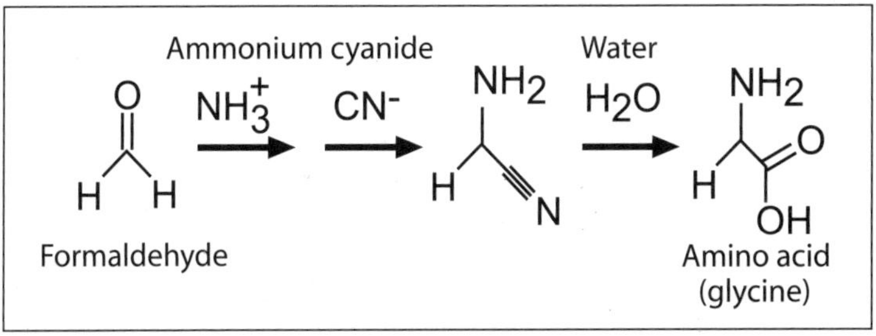

Figure 6.1. Strecker synthesis of amino acids from minimal materials, as in the Miller-Urey experiment. Replacement of one of the hydrogens of formaldehyde with other atoms leads to other, more complex amino acids instead of the simplest one, glycine.

environment is anaerobic and poisonous by modern standards, aldehydes (like formaldehyde) and nitriles (like cyanide) being quite toxic to creatures evolved in a world in which metabolism has adapted to global oxygenation. This is one reason why we cannot easily study abiogenesis today: the world that could host it is gone—and good riddance.

Of the three signs of the chemical origins of life emphasized here, the final one is that we are not completely dependent on earthly processes for the monomers of proteins and the bases of nucleic acids. Formaldehyde (and more complex aldehydes) and hydrogen cyanide (and more complex nitriles) are seen in the light from the solar nebula. They are, in fact, among the most prevalent molecules in the interstellar space of the Milky Way galaxy. Furthermore, no guessing is required about the relation of this observation, based on light from space, to an early Earth. In 1969 a carbon-containing remnant of the early solar system hit the Earth: a meteorite known as the Murchison carbonaceous chondrite. When it was analyzed it was found to contain several percent of its carbon as amino acids like those seen in the Miller-Urey experiments and bases like those studied by Oró. We surely are stardust—not only were our "heavy" atoms (e.g., carbon and nitrogen) made in ancient stellar supernovae, but the ingredients for our multiatomic genetic molecules are also spread between the stars. The implications of these facts range outward from the Earth. Life might be relatively frequent in this neighborhood of the galaxy, if the presence of the basic bits needed for its synthesis is taken as evidence. Life's substance may ride the wind that blows between the stars.

So why is there raging and indefinite controversy about abiogenesis? Important biomolecules will occur spontaneously when interstellar materials are hammered by some source of energy to promote rearrangement. If all we need is a primordial replicator to kick things off, the game of living seems to be virtually afoot.

But though nucleobases are easy to make, putting together nucleic acids that might encode information and replicate has so far proven impossible. Whereas sugars arise with ease from another of the great mother molecules (formaldehyde polymerizes to sugars in slightly alkaline water), the ribose sugars of RNA and (even more so) the deoxyribose sugars of DNA are minorities among the products. Furthermore, it is difficult or impossible to put preformed sugars and bases together to create sugar-base nucleosides for a nucleic acid, and difficult again to put the phosphate on the resulting molecules to yield nucleotides.

The inability to synthesize nucleotides and nucleosides has endured so long that it has sometimes seemed an insurmountable barrier. However, an easy, efficient route to U and C nucleotides that uses simple, plausibly ancient chemicals has turned up. The new synthesis came from John Sutherland's laboratory in Manchester; it was developed thanks to rigorous focus on the desired products, creative exploration of possible reactions, and simultaneous use of the same molecules to join the products and also direct intermediate reactions. In these reactions, a crucial intermediate compound yields pieces of both base and sugar, thereby avoiding the difficult step of joining premade base and sugar chunks. Such developments suggest that improved, higher-yielding methods for the production of nucleotides for RNA may lie ahead.

However, it is still difficult to assemble nucleotides into chainlike RNA and DNA polymers that compose messages and carry out reactions. Moreover, even the reactions that we know how to carry out (recall Stanley Miller's spark lightning, and the use of freezing to stimulate the Oró reaction) seem dreams from different worlds, not necessarily concrete possibilities from our planet's past.

There are two main scientific reactions to these broadly acknowledged facts. The first is that there was a pre-RNA world whose inhabitants actually invented replication, and therefore Darwinian evolution. These hypothetical pre-RNAs (imagined to take forms as different as the pattern of defects on a clay surface, or as simplified RNA molecules with other backbones) are the ancestors whose components were easily made and assembled into replicators by Earth's Hadean chemistry. The highest accomplishments of these Hadean ancients were to pave the way for RNAs by performing the synthetic reactions that lightning and cold cannot easily do alone.

However, there is a second possibility, one that I strongly favor. I have no problem believing that RNA was not the first replicator, the brilliant originator of the Earth's Darwinian era. This book is in fact probably not the story of the origin of life, but of a more diverse golden age thereafter—the RNA world. It is difficult, though, to be specific about another genetic molecule before nucleic acids. Fossil clues about these hypothetical pre-RNAs might well be expected to persist among the molecules of modern cells, but they are not apparent. Perhaps these ancient leftovers from two worlds ago are hiding inside us in plain sight. It is a diverting exercise in detective work to thumb through a biochemistry book, questioning all the conceivable subjects, searching for them.

But the second possibility worth considering is that there might have been a special environment, yet to be envisioned, in which a small replicator ultimately related to RNA was a frequent natural product.

In fact, the continuity of evolution (or one of its logical equivalents) provides us some help in envisioning the pre-RNA—and the pre-pre-RNA as well, if it comes to that. Continuity is an essentially mathematical idea. If you were drawing a graph of a continuous function, you could draw a smooth line. A discontinuous function has a graph that goes shooting off the top or the bottom of the page at the discontinuity (say, because you are forced to divide by zero at that point). To put this in terms more relevant to evolution, evolution does not jump, but must have closed in on and exited smoothly from every door we know it used.

This turn of phrase has its limits, because unlike the continuous line on a graph, evolution actually proceeds by the finite jumps called mutations. If humans have about 27,000 genes, then, speaking approximately, when they evolve they usually change in a genetic or evolutionary sense by a few parts in 27,000, which is a small but finite alteration. And it can be an essential one, if one of the altered parts has a critical job, so that a mutation causes a genetic disease and thus makes the body unworkable.

But setting the finite digital character of genomic texts aside, evolution makes small changes. Not only are the most likely mutations small, but those mutations that prove successful in evolution are even more surely the tiny ones. Large changes are unlikely to fit an organism to its circumstances; instead they are likely to jump off the board and take the organism out of the game, straight to extinction. Before the full elaboration of Darwinism, the idea that organisms sometimes made large but successful genomic leaps—so-called "hopeful monsters"—was debated, but more recently such mutations have been found to be quite rare. The changes that benefit a creature are almost always small: modern Clydesdales evolve from tiny ancestral horses by an almost-continuous series of small steps.

The point with respect to abiogenesis is this: the only way to easily make large molecules is to link together threads of small, identical molecules. This remains the dominant biological path to informational polymers today. (See Chapter 10 for an example.) It is a much simpler method than growing through many chemically diverse additions, because it repeats the same successful, well-known trick over and over. This approach to synthesis implies in turn that the initial replicators will elaborate into long, thin molecules with repetitive monomer structures—bumpy threads like DNA, RNA, and unfolded proteins.

The only simple way to make such extensive chains is to use carbon chemistry. We therefore would expect the earliest replicator, among all reproducing forms, to be the carbon compound most like the primordial environment. This is required because the earliest replicator emerged from a nonliving collection of organic and inorganic chemicals through a small, practical modification. Its chemical innovation, though necessarily minor (obeying the principle of continuity), is crucial. The modified molecule, the initial replicator, becomes quasi-immortal because it will evolve to be better and better at propagating itself. More efficient capture of its own building blocks will automatically lead to more abundant descendants. With an increase in size also comes the ability to embody better instructions for its own replication. Words of two letters are more expressive and more accurate than words of one letter, and so on.

What, then, is this simplest replicator, rooted directly in carbon geochemistry? Is it RNA-like or very different (though continuously connected to RNA by descent), given that it could emerge in a probable environment on the Hadean Earth?

Our thoughts are led to chemically and physically simple replicators, containing units that can evolve to become nucleic acid monomers, as the center of the origin-of-life discussion. Furthermore, this idea presents an explicit experimental challenge, and exploration of it should yield real data in a real laboratory. While there is as yet no consensus as to the identity of the breakthrough replicator, I have included a guide to experiments on simple chemical replicators in the suggested readings for this chapter.

But if you should undertake this experimental quest, keep in mind that origin time was about 4 Gya. The very rocks of that era have for the most part perished, long since subducted or metamorphosed, taking any possibility of a pre-RNA fossil with them. It is difficult to visualize a world so distant in time, and to mentally survey environments that might have nurtured the synthesis of complex chemicals. For example, it is possible to combine the electric fire of the Miller-Urey reactions and the ice of the Oró reaction. Where were the Hadean cold places (apparent oxymoron though this may be; see Figure 5.1) where the two could coexist?

Nevertheless, natural synthesis of a primordial replicator remains a hypothesis worth pursuing, through every subterranean, atmospheric, and oceanic environment that emits a glimmer of relevance.

Readings

"Organic compound synthesis on the primitive Earth." Stanley L. Miller and Harold C. Urey. *Science* 130: 245–251 (1959).
The classic—still interesting reading 50 years later.

"The origin of life on the Earth." Leslie E. Orgel. *Scientific American* 271: 76–83 (1994).
A useful review of life's beginnings from an eminent worker in origins chemistry.

"Synthesis of purines under possible primitive Earth conditions. I. Adenine from hydrogen cyanide." J. Oró and A. P. Kimball. *Archives of Biochemistry and Biophysics* 94: 217–227 (1961).
The demonstration that there is a plausible cosmic route to nucleobases.

"Minimal self-replicating systems." Natasha Paul and Gerald F. Joyce. *Current Opinion in Chemical Biology* 8: 634–639 (2004).
A short review that will lead you to descriptions of simplified replicating systems based on nucleic acids, based on organic chemicals that emulate nucleic acids in order to replicate, based on ordinary organic chemicals, and even based on peptides that, with elaborate preparations, replicate.

"Synthesis of activated pyrimidine ribonucleotides in prebiotically plausible conditions." M. W. Powner, B. Gerland, and J. D. Sutherland. *Nature* 459: 239–242 (2009).
The chemical accessibility of C and U nucleotides is obvious only after the route is found, as is well illustrated here. This paper is probably the hardest reading in this list, but worth the effort for anyone with a knowledge of elementary organic chemistry. The reader's reward is an exceptional example of focused and creative experimentation to ultimately elucidate a classical problem.

"Borate minerals stabilize ribose." A. Ricardo, M. A. Carrigan, A. N. Olcott, and S. A. Benner. *Science* 303: 196 (2004).
A vivid example of help from special environments. Borate is a mineral with a stable interaction with ribose. Thus borate substantially tames the chaos of natural undirected sugar synthesis, favoring five-carbon sugars, including RNA's ribose.

7

Tornados in a Junkyard

Blotherasts argle contornaceously bethwart mungled chardwicks and fintipled mesterlinks.

—Nonsense by Noam Chomsky

A junkyard contains all the bits and pieces of a Boeing 747, dismembered and in disarray. A whirlwind happens to blow through the yard. What is the chance that after its passage a fully assembled 747, ready to fly, will be found standing there? So small as to be negligible, even if a tornado were to blow through enough junkyards to fill the whole Universe.

—Fred Hoyle, *The Intelligent Universe* (1983)

We share a common ancestry with our fellow primates; and going still further back, we share a common ancestry with all other living creatures and plants down to the simplest microbe. Darwin's theory, which is now accepted without dissent, is the cornerstone of modern biology. Our own links with the simplest forms of microbial life are well-nigh proven.

—Fred Hoyle and Chandra Wickramasinghe, *Lifecloud: The Origin of Life in the Universe* (1978)

We now follow the starry winds to Earth, where they become Hoyle tornados. Astronomer Fred Hoyle named the Big Bang as a derisive joke. Yet the joke backfired when the bang became the commonly accepted explanation for the universe's origin. In fact Hoyle would have favored an alternative steady-state model, in which the universe is always and continuously created. But though he would have preferred a non-evolving cosmos, Hoyle was not an opponent of Darwinian evolution, as the second quote from him demonstrates.

Nevertheless, the risible, striking Hoyle image of an airliner assembled by the arbitrary whirling of junked parts, ready for takeoff, has entered the imaginations of Darwinians and doubters alike. It lives on in many a turn of phrase and book title, and I allude to it because it captures a serious argument that must be answered. Are living creatures too complicated to have arisen except by the directed action of an intelligent creator? In fact, this argument holds appeal even for some physical scientists. A decent respect for the opinions of mankind requires us to confront it.

The Hoyle tornado is not even close to an appropriate model of evolution, which never does and never can put together a complicated thing with one blow. Instead, evolution cobbles together something that works in many small, hemicontinuous intermediate steps, to make a gradually more useful machine. The Hoyle tornado is instead more relevant to abiogenesis, the advent of life (see Chapter 6). But let us leave this protest aside in order to embrace a more widely occurring form of the tornado, which turns not on whether evolution can leap to a conclusion, but on the complexity of the result.

An early form of the argument sets a tone that echoes down the decades to yesterday's discussions. William Paley, an eighteenth-century theologian, first set the watchmaker's watch in 1802: "Suppose I had found a watch upon the ground, and it should be inquired how the watch happened to be in that place." To Paley, the intricate functionality of the watch insists "that the watch must have had a maker . . . who comprehended its construction, and designed its use." So also for Paley and many others since, life must owe its complexity to a designer. This notion survives as perhaps the most frequently offered of the arguments that Darwinian evolution "cannot work." We will consider it in more detail, but for the moment note that it requires some assumptions—watches and lizards are similar, and I "cannot imagine" that they do not share qualities of origin. Such arguments from personal disbelief are particularly vulnerable to new information, as I hope we will see.

An interesting sidelight is that complexity itself can be measured. So complexity has attracted considerable attention from scientists themselves, particularly physicists and chemists (like Hoyle), who are accustomed to thinking about elaborate systems—systems with vast numbers of possible states. I have a colleague in the "hard" sciences at my university who believes that the complexity argument mandates the "impossibility" of Darwinian change. Of course, it is not Darwinian change that is under discussion here, but abiogenesis, the appearance of life, a more difficult topic. Nevertheless, listen to Harold Morowitz, a biophysicist now at George Mason University. In *Energy Flow in Biology* (1968), he writes that if we take a flask of bacteria and heat it up to destroy its highly ordered structure and reduce it to an equilibrium mixture of chemicals (chemicals that necessarily should be appropriate for the construction of living things), the probability that a bacterium would be among the products resting quietly at equilibrium is 1 in

$10^{100,000,000,000}$!

Yet this calculation (leaving aside certain mathematical reservations) not only asks for life in one step, it asks it to form at chemical equilibrium—that is, without sunlight, without energy input of any kind. This particularly strenuous variant of spontaneous generation has to be improbable, and so it is.

Both Hoyle and Morowitz are scientists and rationalists trying to grapple with life's intricacy. Living things are complicated, and this is certainly worth thinking about. The modern intellectual descendants of William Paley are not Hoyle, Morowitz, and friends but the enthusiasts of intelligent design. To examine their argument, we accept the idea that living things are complex and

Numbers with exponents

The exponential presentation of numbers used in scientific conversations and in the text is not difficult to understand. Furthermore, it is useful to understand exponential numbers because it makes it easy to discuss things that are vastly different—in size or likelihood, for example.

Here is the basic trick: numbers are simplified by collapsing the number of zeros they have into a new part of the number, a superscript called an exponent. For example, instead of writing one thousand as 1,000, write 10^3. Here we are saving one digit by using 10^3 for 1,000 ($= 10 \times 10 \times 10$, where the 3 stands in for the 3 10s that are multiplied). 3,000 is 3×10^3. This is all that is required at the first level of the exponent trick; if you want to go no further than thinking of the first number followed by the number of zeros in the exponent, you can read the numbers.

You are probably underwhelmed by the brilliance of exponents so far, but now suppose you wish to discuss an American trillion (the order of magnitude of the dollar cost of a war or a national economic purposeful, and we consider whether they must therefore represent the workings of a synthetic intelligence.

For example, William Dembski, a mathematical philosopher at the Discovery Institute in Seattle (an intelligent design think tank), argues in *No Free Lunch* (2001) that Darwinian mechanisms cannot generate the "specified complexity" that we see everywhere in living things. Dembski sees no route to the kind of complexity that is directed to the solution of a particular problem. In his own words, "any specified event of probability less than 1 in 10^{150} will remain improbable even after all conceivable probabilistic resources from the observ-

bailout), 1,000,000,000,000. 10^{12}, the same number exponentiated, is much easier to handle. But the trick gets even better. You multiply numbers just by adding their (usually small) exponents: $100 \times 1000 = 10^2 \times 10^3 = 10^5 = 100{,}000$. Or better yet, $10^{12} \times 10^2 = 10^{14}$. You can also divide, by subtracting exponents: $10^{14} \div 10^{12} = 10^2$, which we recognize as correct computation because of the way we previously created 10^{14}. Three trillion is now 3×10^{12}, and multiplication and division involve doing the relevant operations on both parts of the number: $3 \times 10^{12} \times 2 \times 10^2 = 6 \times 10^{14}$.

Now for small numbers: 6×10^{-14} is 6 divided by 1-followed-by-14-zeros. Everything else, as they say in textbooks, is left as an exercise for the reader.

But keep in mind the basic rule: if you just think of any exponent as the number of zeros following the number (when the exponent is positive) or the number of zeros in the divisor (if the exponent is negative), you have nearly everything.

able universe have been factored in. . . . It follows that if such systems are specified, then they are designed." William Paley's watch is clearly still ticking.

In the end, Paley and Dembski and the various advocates of watchmaker or design-based theories are far from the mark, but their error is a subtle one. In fact, the process of evolutionary change has somewhat the character of the birthday paradox: given 23 people in a room, two of them will have the same birthday about half the time. The true answer surprises us because it emerges from a hidden place in the problem.

To suggest how Darwinian evolution can surf across supposed oceans of improbability, I have set up a mutation-selection demonstration. The exercise is based on an original devised by Richard Dawkins in *The Blind Watchmaker* (1986), but I have instead used an implementation by Rob Knight and Steve Freeland, two research evolutionary biologists. (The quoted aphorism, "Nothing in biology makes sense except in the light of evolution" belongs to Theodosius Dobzhansky, a celebrated fruit fly geneticist and a notable architect of twentieth-century Darwinism.)

To fully appreciate this particular tornado, you must be aware that it is an idealization, not a realistic model of evolution. But its power comes from its precise aim—directly at the heart of the difficulty many people have with the concept of evolution. Think about how hard it is to evolve Dobzhansky's particular saying "Nothing in biology. . . ." We spell it out using the 26 American letters plus a space, and we place these randomly in a string of 63 symbols. The stunning fact is that there exist $27^{63} = 1.5 \times 10^{90}$ such possible strings, or 1 500 000 000 000 000 000 000 000 000 000 000 000 000 000 000 000 000 000 000 000 000 000 000 000 000 000 000 000 000 000 000 000 "statements" we might make in these 63 spaces—many, many more than the number of elementary particles in all the known universe. So finding any one statement seems an unimaginable task.

Or, as a creationist, intelligent designer, or skeptic might say:

> If you tried a million strings a second, starting at the moment of the big bang, you would just now be looking at your 500 000 000 000 000 000 000 000th statement of 63 characters. Unfortunately for your argument in favor of evolution, you have hardly started; you still have fully 1.5×10^{90} strings to examine. The stars in the sky will burn out and your protons will decay before you can get through all the possibilities. If you can't evolve a statement containing only 63 letters, how do you expect to evolve an Earth full of creatures, each of which has millions or billions of characters like this in its genome?

To answer this Paley-Hoyle jibe, we let a computer write out a random string, mutate 1 in 100 characters in each generation, and select changes only if they match Dobzhansky. Here is a typical result:

Round	No. misses	**63-character text, including spaces**
0	55	xabqwjknnejflwlsd ymykwqsoxrsgrkodeqqjcfxtot yold npjloxylqpof
1	54	xabqwjknnejflwlsd ymykeqsoxrsgrkodeqqjcfxtot yold npjloxylqpof
2	53	xabqwjkniejflwlsd ymykeqsoxrsgrkodeqqjcfxtot yold npjloxylqpof
3	52	xabhwjkniejflwlsd ymykeqsoxrsgrkodeqqjcfxtot yold npjloxylqpof
4	51	xabhwjknie flwlsd ymykeqsoxrsgrkodeqqjcfxtot yold npjloxylqpof
5	50	labhwjknie flwlsd ymykeqsoxrsgrkodeqqjcfxtot yold npjloxolqpof
6	49	labhwjknie flwlsd ymykeqsoxrsg kodeqqjcfxtot yold npjloxolqpof
7	48	labhwjknie flwlod jmykeqsoxrsg kodeqqjcfxtot yold npjloxolqpof
8	47	labhwjknie flblod jmykeqsoxrsg kodeqqjcfxtot yold npjloxolupof
9	47	abhwjknie flblod jmykeqsoxrsg kodeqqjcfxtot yold npjloxolupof
10	46	abhwjknie flblod jmykeqsoxrsg koceqqjcfxtot yold npjloxolupof
11	45	abhwjgnie flblod jmykeqsoxrsg koceqqjcfxtot yold npjloxolupof
12	45	abhjjgnie flblod jmykeqsoxrsg koceqqjcfxtot yold npjloxolupofb
13	44	abhjjgnie blblod jmykeqsoxrsg kocegqjcfxtot aold npjloxolupofb

```
14 43 abhjjgnie blblod jmykeqsoxrsy kxcegqjcfxtot aold npjloxolupofb
15 42 abhjjgnie blblod jmykeqsoxrsy kxcegqjcfxtot aold nojloxolupofb
16 41 abhjjgnie blblod jmykeqsoxrsy kxcegqjcfxtot aold nojloxolupofn
17 40 abhjjgnie blblod jmakeqsoxrsy kxcegqjcfxtot aold nojloxolupofn
18 39 abhjjgnie blblod jmakeqssxrsy kxcegqjcfxtot aold nojloxolupofn
19 38 abhjjgnie blblod jmakeqssxrsy kxcegqjcf tot aold nojloxolupofn
20 37 abhjjgnie blblod jmakeqssxrsy kxcegqjcf tot aold nojloxolupoon
21 36 abhjjgnie blblod jmakeqssxrsy kxcegqjcn tot aold nojloxolupoon
22 35 nabhjjgnie blblod jmakeqssxrsy kxcegqjcn tot aold nojloxolupoon
23 34 nabhjjgnin blblod jmakeqssxrsy kxcegqjcn tot aold nojloxolupoon
24 33 nabhjjgnin blblod makeqssxrsy kxcegqjcn tot aold nojloxolupoon
25 31 nabhjjgnin blblod makeqssxrsy kxcegqjcn tot aold nojlovolutoon
26 30 nabhjjgnin blblod makesssxrsy kxcegqjcn tot aold nojlovolutoon
27 28 nabhjjgnin blblod makesssxrsy kxcegtjcn toe aold nojlovolutoon
28 27 nabhjjgnin blblod makesssxrsy kxcegtjcn toe aild nojlovolutoon
29 26 nabhjjgnin blblod makesssxrsy kxcegtjcn toe aild ojlovolutoon
30 25 nabhjjgnin blblod makesssxrsy kxcegtjcn toe aild ojlovolution
31 25 nabhjfgnin blblod makesssxrsy kxcegtjcn toe aild ojlovolution
32 24 nabhjfgnin blblod makesssxrse kxcegtjcn toe aild ojlovolution
33 23 nabhjfgnin blblod makesssxrse kxcegt cn toe aild ojlovolution
34 23 nabhjfgjin blblod makesssxrse kxcegt cn toe aild ojlovolution
35 22 nabkifgjin blblod makes sxrse kxcegt cn toe aildj ojlovolution
36 22 nabgifgjin blblod makes sxrse kxcegt cn toe aildj ojlovolution
37 21 nabgifgjin biblod makes sxdse kxcegt cn toe aildj ojlovolution
38 20 nabgifgjin biblod makes sxdse kxcegt cn toe ailhj ojlovolution
39 19 nabgifgjin biblod makes sxdse kxcegt cn toe ailhj ojlevolution
40 18 nabgiegjin biblod makes sxdse kxcegt cn toe ailht ojlevolution
41 17 nabhiegjin biblod makes sxdse kxcegt cn toe ailht ojlevolution
42 17 nabhieggin biblod makes sxdse kxcegt cn toe ailht ojlevolution
43 17 nabhieggin biblod makes sxdse kxcegt cn toe ailht oalevolution
44 16 nabhieggin biolod makes sxdse kxcegt cn toe ailht oalevolution
45 15 nabhinggin biolod makes sxdse kxcegt cn toe ailht oalevolution
46 14 nabhinggin bioloc makes sxdse kxcegt cn toe ailht oflevolution
47 14 nabhinggin bioloc makes s dse kxcegt cn toe ailht oflevolution
48 13 nabhinggin bioloc makes s dse kxcegt in toe ailht oflevolution
49 12 nabhinggin bioloc makes s dse kxcegt in the ailht oflevolution
50 12 nabhinggin bioloa makes s dse kxcegt in the ailht oflevolution
51 12 nabhinggin bioloa makes s dse cxcegt in the ailht oflevolution
52 11 nabhinggin bioloay makes s nsejcxcegt in the ailht ofxevolution
53 11 nabhinggin bioloay makes s nseicxcegt in the ailht ofxevolution
54 11 nabhingdin bioloay makes s nseicxcegt in the ailht ofxevolution
55 11 nabhingdin bioloay makes s nsebcxcegt in the ailht ofxevolution
56 10 nabhingdin bioloay makes s nsebexcegt in the ailht ofxevolution
57 10 nabhingdin bioloay makes s nsebexceet in the ailht ofxevolution
```

```
 58  9  nabhingdin bioloay makes s nsebexceet in the lilht ofxevolution
 59  9  nabhingdin bioloay makes s nsebexceet in the lilht ofxevolution
 60  9  nabhingdin bioloay makes s nsebexceet in the lilht oftevolution
 61  9  na hingdin bioloay makes s nsebexceet in the lilht oftevolution
 62  8  na hingdin biology makes s nsebexceet in the lilht oftevolution
 63  8  na hingdin biology makes s nsebexceet in the lilht oftevolution
 64  7  na hingdin biology makes s nsebexcept in the lilht oftevolution
 65  7  na hingdin biology makes s nsebexcept in the lilht oftevolution
 66  7  na hingcin biology makes s nsebexcept in the lilht oftevolution
 67  7  na hingcin biology makes s nsebexcept in the lilht ofdevolution
 68  7  na hingcin biology makes s nsebexcept in the li ht ofievolution
 69  7  na hingcin biology makes s nsebexcept in the li ht ofievolution
 70  6  na hingcin biology makes s nse except in the li ht ofievolution
 71  5  na hing in biology makes s nse except in the li ht ofievolution
 72  5  na hing in biology makes s nse except in the li ht ofievolution
 73  5  na hing in biology makes s nse except in the li ht ofievolution
 74  4  na hing in biology makes sense except in the li ht ofievolution
 75  4  n hing in biology makes sense except in the li ht ofievolution
 76  3  n hing in biology makes sense except in the li ht of evolution
 77  3  n hing in biology makes sense except in the li ht of evolution
 78  3  n hing in biology makes sense except in the li ht of evolution
 79  2  n thing in biology makes sense except in the li ht of evolution
 80  2  n thing in biology makes sense except in the li ht of evolution
 81  2  n thing in biology makes sense except in the li ht of evolution
 82  2  n thing in biology makes sense except in the li ht of evolution
 83  2  n thing in biology makes sense except in the li ht of evolution
 84  2  n thing in biology makes sense except in the li ht of evolution
 85  2  n thing in biology makes sense except in the li ht of evolution
 86  2  n thing in biology makes sense except in the li ht of evolution
 87  2  n thing in biology makes sense except in the li ht of evolution
 88  2  n thing in biology makes sense except in the li ht of evolution
 89  2  n thing in biology makes sense except in the li ht of evolution
 90  2  n thing in biology makes sense except in the li ht of evolution
 91  2  n thing in biology makes sense except in the li ht of evolution
 92  2  n thing in biology makes sense except in the li ht of evolution
 93  1  nothing in biology makes sense except in the li ht of evolution
 94  1  nothing in biology makes sense except in the li ht of evolution
 95  1  nothing in biology makes sense except in the li ht of evolution
 96  1  nothing in biology makes sense except in the li ht of evolution
 97  1  nothing in biology makes sense except in the li ht of evolution
 98  1  nothing in biology makes sense except in the li ht of evolution
 99  1  nothing in biology makes sense except in the li ht of evolution
100  1  nothing in biology makes sense except in the li ht of evolution
101  1  nothing in biology makes sense except in the li ht of evolution
```

```
102  1  nothing in biology makes sense except in the li ht of evolution
103  1  nothing in biology makes sense except in the li ht of evolution
104  1  nothing in biology makes sense except in the li ht of evolution
105  1  nothing in biology makes sense except in the li ht of evolution
106  1  nothing in biology makes sense except in the li ht of evolution
107  1  nothing in biology makes sense except in the li ht of evolution
108  1  nothing in biology makes sense except in the li ht of evolution
109  1  nothing in biology makes sense except in the li ht of evolution
110  1  nothing in biology makes sense except in the li ht of evolution
111  1  nothing in biology makes sense except in the li ht of evolution
112  1  nothing in biology makes sense except in the li ht of evolution
113  1  nothing in biology makes sense except in the li ht of evolution
114  1  nothing in biology makes sense except in the li ht of evolution
115  1  nothing in biology makes sense except in the li ht of evolution
116  1  nothing in biology makes sense except in the li ht of evolution
117  1  nothing in biology makes sense except in the li ht of evolution
118  1  nothing in biology makes sense except in the li ht of evolution
119  1  nothing in biology makes sense except in the li ht of evolution
120  0  nothing in biology makes sense except in the light of evolution
```

What has happened here? Somehow random mutation, and selection for something that "works," has cut through the mind-warping improbability of finding any one string out of the 1.5×10^{90}. (Even the "random" string we began with—which we of course found quite easily—is one of 10^{90}.) And it did not require longer than the life of our universe. In fact, it took only 120 steps and a couple of seconds on a laptop computer.

How did we manage to go about 10^{90} times as fast as intelligent design critics suggest should be possible? The secret—as you have already realized—is in the selection, which winnows the random mutations. In particular, we have no need to, cannot, and do not take the time that would be required to visit every one of the possible states of our system of letters. Instead, we head "toward" the functional statement by accepting successively better approximations to it. This simplifying directional tendency makes our otherwise remote goal startlingly accessible. Given only that there are mutational intermediates from which we can better "see" the solution (and that can be selected—a nontrivial assumption), we could indeed evolve a far longer statement.

As a case in point, what would we need to exceed intelligent designer Dembski's "absolute" limit of 1 in 10^{150}? In fact, any statement of 105 characters or spaces or longer lies over his line. For example, "I am gravely disturbed by the capacity of mutation and selection to search large numbers of possibilities" fills exactly 105 places, so its evolution would violate the "Dembski limit." As much as I would like to see those exact words materialize on this page, the supposedly more-than-impossible statement will not fit between our margins, so it is hard to present clearly here. But it can be evolved by the very same method.

People who wish to take refuge in complexity sometimes claim that this kind of result is faked because we included the target statement in our program and then intelligently chose the intermediates. This objection gets the argument backward. Even our starting statement, as indicated earlier, is one out of 1.5×10^{90}. Or, to put it in other words, by induction we get a similar result no matter which statement we pick as the target. The Chomsky gibberish on the opening page of this chapter—written long before and in complete disregard of our present purposes (in order to faithfully emulate English text, but without any meaning)—is one of the statements of our Dobzhansky system and accordingly could be evolved if we so chose.

The point is not that we evolved any one statement but that we might choose any target with similar results—descent with modification and selection clearly will navigate to any and all of the 1.5×10^{90} strings of characters in our system, our toy genome. The objection that we picked a particular goal cannot be to the point, because any possibility would have given the same result, whether or not it is predefined from our point of view.

Furthermore, we can eliminate human choice as a factor: the first English word surrounded by spaces occurs quickly, at the 19th step. We could evolve words and compose sentences without any initial target instead of targeting Dobzhansky's aphorism. Mutation and selection achieve adaptive texts (texts that make sense) without any target.

Alternatively, this creation-design objection may mean that picking one explicit statement is not how evolution works. But this objection is barking up entirely the wrong tree, since real proteins or nucleic acids remain functional if some of the letters are changed. Real biological proteins and nucleic acids are not unique statements but clouds of similar texts. This makes them easier to find than Dobzhansky's singular apothegm, because our path need only hit the cloud of usable statements rather than a unique target.

Make no mistake: evolving Dobzhansky's saying is not a completely realistic simulation of evolution. Back-mutation, reproduction, and other evolutionary fundamentals have been left out. However, the existence of *any* credible route through such vast spaces, rapidly zeroing in on a goal seemingly inevitably lost among myriads, should give intelligent designers (and the rest of us) a reflective moment or two.

But there is another way to make the design argument which has a uniquely ironic flavor—it is an evolutionary critique of Darwinism. William Dembski's colleague at the Discovery Institute, Michael Behe, is a biochemist also associated with Lehigh University. Behe has a particular embodiment of Dembski's specified complexity in mind. In *Darwin's Black Box* (1996), he argues that cellular molecular machines (the bacterial flagellum, the biochemistry of vision) are "irreducibly complex."

"An irreducibly complex object will be composed of several parts, all of which contribute to the function . . . all the components are required for the function." Loss of any part renders the irreducibly complex machine nonfunctional. Thus there could, according to the Behe argument, never be the simpler evolutionary intermediates required for a Darwinian path to complex molecular machines. The necessary intermediates, like those we used to evolve our Dobzhansky quote, could never have existed.

Somewhat surprisingly, Behe believes in the evolution of visible structures, though this interesting view does not usually appear in discussions of intelligent design. He concedes that drug-resistant bacteria, or the beaks of Darwin's finches, or moths whose coloring darkens over time after their perches in trees turn dark from industrial soot, are behaving just as Darwin predicted. He therefore concedes that Darwin's idea has "triumphed"—except when molecules are considered. One wonders where exactly to place the boundary at which Darwinian triumph turns to rout—evidently somewhere between organisms of bacterial size and molecular complexes. Most believers in intelligent design do not seem to know that Behe has already conceded everything bacterial and larger to Darwinian materialism.

To illustrate irreducible complexity, Behe cites a spring-driven mousetrap, which cannot catch a mouse if any part (the catch, the spring, the platform) is omitted. But the Web site of John McDonald of the University of Delaware is particularly interesting in this light—it shows that the complexity of a mousetrap is in fact, despite Behe's assertion, reducible. McDonald's traps begin with a piece of the trap's spring wire alone as an easily triggered snare, bent on itself and sensitively poised so that it can spring closed if touched by a passing mouse. It does no good for intelligent design advocates to object that McDonald's creations do not show how the mousetrap can evolve; no one has claimed that it evolved. But McDonald's mousetraps beautifully exhibit the hazards of arguing from inconceivability—once someone has shown how the mousetrap could have once been a simpler device, the irreducible has been reduced—and there is no going back.

Perhaps the eye, which did evolve, is a better example. It certainly comes up frequently in arguments. Is the mammalian eye irreducibly complex, so that a part of the eye would by itself not be useful? It would not seem so: the detection of any light at all by a mere flat, light-sensitive eyespot would enable you to tell down from up during daytime, or to detect a potential predator between you and the sky. In fact, Dan Nilsson and Susanne Pelger of Lund University have modeled the progress of eyes evolving from such an eyespot, and they find that there is a smooth series of small changes that could be selected for increased acuity. For example, starting with a light-sensitive spot, a cuplike eyespot gains directionality, a fluid-filled cup gains focus, and so on. With some plausible, perhaps pessimistic, assumptions about the rate of evolutionary change, one could evolve a small, mammalian, camera-like eye in a few hundred thousand years. This is about 1/1,000 to 1/2,000 of the time that we have had since complex cellular body plans appeared on Earth, so Darwinian routes to an eye seem quite plausible. This conclusion is confirmed by the many independent appearances of eyes, as counted up by evolutionist Richard Dawkins in a 1995 article in *New Statesman and Society:*

> Serviceable image-forming eyes have evolved between 40 and 60 times, independently from scratch, in many different invertebrate groups. Among these 40-plus independent evolutions, at least nine distinct design principles have been discovered, including pinhole eyes, two kinds of camera-lens eyes, curved-reflector ("satellite dish") eyes, and several kinds of compound eyes.

There are more philosophical objections to design; because a designer must comprehend the information necessary for the object to be designed, the designer must be more complex than the designed object. So how then did the designer arise? Sometimes this gambit is declined by intelligent designers who explicitly acknowledge that they are thinking of the Islamo-Judeo-Christian God. Without this admission, every creation recedes into an infinite regression of designers. Design therefore creates a problem for itself—it cannot avoid posing a problem greater then the one it sets out to solve.

A final point is that there are often simple ways to reach what, looking back, appears to be irreducible complexity. A famous example is the stone arch. Imagine an arch built on top of wooden scaffolding. The scaffold holds the stone in place until a complete, self-supporting arch is completed by placement of the keystone. Now take the scaffold away. All the archstones are now required; the finished arch is "irreducibly complex," collapsing if any part is taken away. Yet it will not be correct to argue that some mysterious force must have supported the archstones before they were all in place. In fact, the now-discarded scaffold, put away once the arch was sufficiently evolved, did the job handily.

However, to talk of evolution we don't necessarily need theoretical arguments. There is an experimental science of evolution played out on laboratory benchtops (compare with Chapter 14). In such experiments we see new structures arise by way of a controlled evolutionary process. Because this is an area in which my laboratory works, I stress that this field of biology has very pointed implications for intelligent design.

The idea, originated about 15 years ago in three separate laboratories (see the end of Chapter 14), is that evolution of a genetic text could be carried out by the techniques of biochemistry. Recall that a nucleic acid is a linear string of four symbols that can be reproduced or replicated. A population of nucleic acid molecules (often RNA) that has a randomized sequence (all four nucleotide letters, A, C, G, and U, at similar positions in different molecules) is subjected to selection. There is mutation, either mild (in the form of normal mistakes of replication) or stronger (by forcing the replication to be error prone), and selection in some form, say, by having the RNA molecules bind to plastic beads whose surface is linked to a small molecular target. You wash the RNAs off the beads by flowing past them a free form of the target molecule. Because there are more free targets than targets bound to the beads, the target-binding RNAs bind the free ones and as a result float off the beads. Bind, wash, replicate. Bind, wash, replicate. In the end, you purify the molecules that work—those that bind to the molecule that decorates the beads. The process emulates Darwinian proceedings relatively accurately.

For each target molecule, after a few cycles you get a new variety of RNA structures that bind to it. For good targets (those that stick easily to RNA), many different binding structures are found. Newly selected binding structures appear to be irreducibly complex: they are inactivated by deletion of or change in single RNA nucleotides. They are apparent examples of specified (functional, binding) complexity arising by Darwinian means.

William Dembski is aware of the existence of experimental evolution, and in *No Free Lunch* he replies that intelligence is required to make RNA and to winnow the mixtures of molecules for their ability to stick to small targets. According to him, the varied functional products of experimental evolution, though real, are just the disguised intelligence of the designer-experimenter.

However, this cannot be a sufficient explanation. The experimenter does the same thing in every experiment: she makes the sticky beads; she turns the nucleic acid replication crank. Yet the outcome, always unknown beforehand, is different in every experiment. In fact, the outcome is often surprising in its variety of molecular shapes and molecular performance. The experimenter's input is much the same in every experiment—these varied, complex, emergent results cannot be coming from the experimenter, even unconsciously. The only element that varies from Darwinian case to Darwinian case is the target itself. Thus the only possible conclusion is that the target molecule determines the outcome. My laboratory, as predicted from this notion, has shown experimentally that a given target indeed robustly gives its own reproducible result if the simplest active structure (instead of a broad variety of structures) is sought.

A note from Michael Behe's colleagues

As an exercise of authorial license, I give the final word to Michael Behe's own Faculty of Biological Sciences at Lehigh University, as quoted on its Web site in 2006: "The department faculty, then, are unequivocal in their support of evolutionary theory, which has its roots in the seminal work of Charles Darwin and has been supported by findings accumulated over 140 years. The sole dissenter from this position, Prof. Michael Behe, is a well-known proponent of `intelligent design.' While we respect Prof. Behe's right to express his views, they are his alone and are in no way endorsed by the department. It is our collective position that intelligent design has no basis in science, has not been tested experimentally, and should not be regarded as scientific."

Intelligent designers might therefore suggest that the input of design or complexity occurs during the synthesis of the target. However, this notion seems to me to concede the evolutionary point, because this benchtop evolutionary process parallels so exactly what happens when organisms evolve, say, to bind and thereby neutralize a toxin in their environments. Experimental evolution is therefore doing what real creatures do. In particular, the hypothetical target antibiotic could be a natural product, made by another microorganism, and therefore not directly designed by any higher intelligence. Or the antibiotic could be even simpler, say, a toxic metal ion from geological sources. In these cases we seem to have specified irreducible complexity arising in the RNA molecule that binds the toxin without any evident place for designer input.

But one final redoubt may remain. Perhaps the intelligent designer intercedes: she designs every antibiotic and RNA, and so continually intervenes in homely processes all over the world. This is exceedingly unlikely. Every scientist making measurements, every engineer calculating performance, every Boeing 787 lined up on the active runway, indeed every computer scientist structuring code to give a desired result is an antenna that could detect such intervention. Thus intervention that changes physical, electrical, chemical. or biochemical outcomes must be rare or nonexistent—unseen in uncounted billions of everyday tests. Instead, as common sense predicts, experiments, calculations, and events with the same input and context produce the same results. Scientists and airline passengers alike are fixated on reproducible outcomes, so this kind of design explanation and this kind of intervention are clearly especially rare or imaginary.

What is the result of all this, including the modern ability to summon evolution onto a designated laboratory benchtop? It seems that novel, complex, functional structures readily arise from the naturalistic processes that science uses to explain the world. It is not being credulous to think that naturalistic processes may explain the diversity seen among the creatures of the Earth. Instead, evolutionary thinking is a simple, plausible reaction to a world of evidence.

Readings

Darwin's Black Box: The Biochemical Challenge to Evolution. Michael Behe. Free Press, New York (1996, 2006).
The quintessential modern inheritor of the Paley argument.

No Free Lunch: Why Specified Complexity Cannot Be Purchased without Intelligence. William Dembski. Rowman and Littlefield, Lanham, Md. (2001, 2006).

A more strenuous read than Behe, because it ranges into mathematically phrased arguments; nevertheless essential for those who need or wish to grasp the authentic voice of intelligent design.

"Understanding the intelligent design creationist movement: Its true nature and goals." Barbara Forrest. http://www.centerforinquiry.net/uploads/attachments/intelligent-design.pdf.
The legal, political, and social context (as of 2007) for the struggle with intelligent design, from an academic philosopher and veteran of the legal battle against its inclusion in school curricula.

"A reducibly complex mousetrap." John McDonald. http://udel.edu/~mcdonald/mousetrap.html
John McDonald's ingenious 2002 demonstration that the spring-driven household mousetrap is not irreducibly complex.

8

Between Genomes and Creatures

We all have genes. This is a universal issue. It is something that impacts everyone.

—U.S. Senator Judd Gregg

Having introduced, back in Chapter 4, the thought that organisms might have descended from simple replicators—chemicals with an unusual knack for organizing their surroundings—we now mull over the idea that they are still chemicals. Gloriously complicated chemicals they may be, but just how much glory and how much chemical? That is our topic.

So just what is an "organism"? There are horses for courses, surely. Sometimes a horse is a mammal; sometimes it is an athlete, supremely quick on its feet. But for many purposes, an organism can be summarized by its genome. Genomes specify the structures of the materials of which organisms are made, directly or indirectly. More interestingly and mysteriously, they also specify a recipe for the assembly of the organism through space and time. This output of genomes is what is seen and altered by evolutionary influences. Today genomes are usually DNA nucleotides, except for a relative handful of RNA viruses (some important to us as human pathogens). We can be sure that genomic information is complete, sufficient to the creature, because it sometimes acts without help. For example, when viral DNA is sent into cells to subvert them into making new viruses, the virus genome acts alone. And the DNA packed within the head of a sperm cell is evidently sufficient to guide the successful development of a complex animal. (Actually, there may be genomic information that accompanies DNA in other molecules, but for the moment we can approximate the genome as consisting of only DNA.)

Our genome is the array of DNA sequences, threaded through the chromosomes of the cell's nucleus, which together contain the recipe for a creature. But just what is in those DNA sequences? We ought to be able to learn something about what it means to be a human from analysis of the recipe for a human. It comes as a shock to realize that biologists are newly uncertain as to the detailed definition of the human genome, or of the genome of any complex (eucaryotic) creature. Simple cases are indeed simply understood; the genome of a bacterium is mostly information for production of its encoded proteins. Surely then our genes for proteins must make up an important fraction of the human genome?

But no—not when judged by abundance. In fact, only about 1.2%, or 1 part in 80, of the human genome is actually the nucleotide code for the linear sequences of amino acids that make up a human cell's proteins. Messenger RNAs for the proteins, made as partial copies of genes, contain untranslated regions (nonprotein information) at both ends and in the middle, of course, but the addition of these brings us to only about 2% of the DNA, or 1 part in 50 of the genome. What are the other 49 parts? Somehow, as we became more complex creatures, we left the genome-as-list-of-parts behind.

Some of the other 98% of human genomic DNA consists of devices made of DNA, to do the things that DNA must. There is a cellular DNA economy, with DNA sequences that direct the marking or sorting or replicating of DNA. There are short sequences (called promoters) marking out the genes as sites from which RNA must be made. There are regions (origins and terminators) that specify tracts of DNA for duplication. There are special DNA sequences at the points where chromosomes are pulled apart during cell division (centromeres) and that serve as protective caps for the ends of chromosomes (telomeres). And there is an unexpectedly large amount of DNA that is the result of accidents—junk from accidental replications and the wreckage from retrovirus invasions of our DNA.

However, although many of these small DNA sequences can be anticipated and named, they still do not adequately account for the other 98% of the genome. Perhaps it is "junk DNA"; yet three-quarters or so of the genome may be transcribed into RNA—a hallmark of consequential genetic expression. Most of this expression has been hidden until recently, because most of these RNAs have unusual termini, and the termini were the handles by which biologists grasped the cell's RNAs when they examined them. Most of these extra RNAs remain inside the nucleus, and it was often also assumed that all interesting RNAs had to be exported from the nucleus to reach the ribosomes, where proteins are made in the cytoplasm.

Understanding the synthesis of proteins from the genes was a triumph of twentieth-century biology. It was an accomplishment not to be minimized, and I in particular am not trying to minimize it here. But we also shouldn't minimize a complementary genomic truth: we cannot yet determine what part of the DNA in a complex organism's genome is meaningful. Nor, on the whole, can we determine what a genome says, or how it says it. For example, information for the synthesis of proteins and information for scheduling might differ in the density of their meanings, and in the language used to express these different genomic purposes. A more complete idea of the genome is coming, bound up in the still-elusive functions of many recently discovered RNAs.

All the same, it is interesting to work out how much might be said in a genome all told—the maximum amount of information stored in a genome—if only as an exercise in grasping the ultimate complexity that has appeared over 4 gigayears of change. So we provisionally take all nucleotides as equivalent—an approximation, but one that allows us to say some things clearly. A genomic nucleotide is about 2 bits of information—a choice between 4 (= 2^2) roughly equally probable alternatives (A, C, G, T/U). Thus a haploid (single maternal or paternal copy) human or mammalian genome is about $2 \times 3 \times 10^9 = 6$ American billion bits. A "fruit fly" is $2 \times 0.12 \times 10^9$ bits = 0.24 billion bits, and our sometimes-friendly colon bacterium *Escherichia coli* is 0.0092 billion bits.

To help us grasp these numbers, I call on our familiarity with the digital information stored in a portable music player—an iPod, Archos, or Zune. How much are 6, 0.24, and 0.0092 billion bits? CD-quality sound is 44,100 samples/second at 16 bits/sample, or 42×10^6 bits/minute. For 2-channel stereo (L and R channels, the usual music format), make it 84 megabits/minute. However, portable music is compressed to save space. Checking my own player, I find that its shrunken music occupies 31 megabits/minute played. Thus we could potentially store just over three hours of reasonably faithful music in the space afforded by a human genome, over 7 minutes in an insect genome, and about 17 seconds in our example of a bacterial genome. Compressed music is unusually densely ordered information, each second split and enumerated in tens of thousands of numbers.

Despite these potential differences between the language of digital music and that of genomes, our comparison sets their variations in a new light. In these terms, a human being is an opera, having comparable complexity and scope for variation. An insect is an aria or a long song; a bacterium is a lyrical line or single musical phrase. Each genome is exquisitely suited to its style of life, just as an opera, a song, or a single musical phrase can, each in its own way, poignantly express something otherwise unutterable. But the complexity of these apt and capable utterances differs greatly, as would be expected from the sizes of their genomess.

Reading

"Relative differences: The myth of 1%." Jon Cohen. *Science* 316: 1836 (2007).

A meditation that neatly illustrates this chapter's subject: it is surely provocative, but what does it really mean to say that human and chimp genomes are only 1% different?

9

A Thumbnail Molecular Biology

There is much pleasure to be gained from useless knowledge.

—Bertrand Russell

A living thing is distinguished from a dead thing by the multiplicity of the changes at any moment taking place in it.

—Herbert Spencer

We need to develop the story of how our subject molecule, RNA, is wound right through the interesting transformations of cells (and noncellular exploiters like viruses). An understanding of this tableau is required for appreciation of the RNA patrimony of modern cells, including our human cells. It turns out that a molecular perspective on cell function immediately and forcefully suggests that the RNA world's contribution to our being has been decisive.

This chapter therefore is intended to provide readers a description that will sum up to an introductory molecular biology. I sketch the standard molecular and informational connections in cells in order to play variations on these themes when we later begin to talk about a previous form of life, the RNA cell or ribocyte. While the names given to our genes and their executors are (like most names) somewhat arbitrary, they are not at all useless in Bertrand Russell's sense; instead they supply a useful patois for describing the molecular furor within a cell.

Those who wish even more detail can turn to the lexicon at the end of the book, where the activities of some cellular RNAs are more formally and specifically defined. On the other hand, those who already feel sufficiently informed can simply jump ahead to Chapter 10, returning here if and when so moved. Or, if you are in the mood for a more modest jump, then skip ahead to the last section of this chapter, which sums up the complete text.

The Molecular Business of Cells

A cell is a device that computes an organism, using the information in the genome and the rest of the cell as its implementers. As intermediates in this computation there are also locally useful copies of parts of the organism's recipe containing both qualitative and quantitative directions (RNAs), as well as peripherals to carry out the instructions in the RNA recipe (ribosomes, enzymes, and enzyme products, such as lipids, carbohydrates, and small molecules). A small genome may compute a microbe, equipped for wandering as atiny but sophisticated single cell that must respond appropriately to the threats and treats in an erratic environment. A longer genome may compute a much larger cell with a more complex program, say, one that changes its surface conductivity to fire electrical spikes when light has fallen on a small retinal field within a mammalian eye. Any cell has a program containing its overall plan, as well as recipes for its participation in more complex organismic parts, such as tissues, when these exist.

Though the plan for an organism is stored as DNA, the execution of the plan involves the transliteration of the DNA plan into RNA, and then the execution of the RNA version of the plan by separate machines, often also with RNAs at their core. We now try to make this fuzzy scheme more specific.

First, let's look at that small but crucial part of the DNA information that encodes protein structures. Then we'll expand these ideas to take in noncoding regions of the genome, which have nothing direct to say about the structure of proteins but serve a myriad of structure and control functions.

Translation Is an RNA Business

Translation, the genome-specified synthesis of ordered chains of amino acids (proteins), is where we begin. Also known as protein biosynthesis, it is the process that connects the sequence of nucleotides and the sequence of amino acids. Not only is translation, in the cytoplasm of an eucaryotic cell, a particular arena in which many cellular RNAs act, but it is believed to be based on a set of reactions once devised and carried out by RNAs alone (see Chapter 16). It is therefore RNA business, first and still foremost.

Proteins are spun out on a digital loom called the ribosome. Ribosomes in turn are folded RNAs (rRNAs) decorated, roughly speaking, at the periphery by 60 or so proteins. Ribosomes differ somewhat in bacteria, archaea, and eucaryotes, but they are recognizably the same kind of machine in all of life's present domains. Ribosomal particles are bipartite, with a smaller section that does the decoding (the selection of amino acids as aminoacyl-tRNAs) and a larger section that joins amino acids carried on the aminoacyl-transfer RNAs (aa-tRNAs). Both ribosomal pieces wriggle together to shuttle aa-tRNAs and the messenger RNA (mRNA), containing the linear code for the amino acids, across the ribosome as it moves down an RNA message. Considerable shuttling (precise repetitive motion, still not understood) is required in order to assemble the hundreds to thousands of amino acids of a protein.

The digital nature of the loom is particularly important; not only does the ribosome chemically link amino acids, but it specifies their order by providing an environment that enhances the accuracy with which aa-tRNA anticodons can base pair with the adjacent specific triplet codons of the mRNA message in order. Not only nucleotide-triplet amino acid signals but also starts and stops are interpreted in this way on the coding (decoding) sites of the ribosome, and some regulatory signals encoded in translational RNAs are executed there.

Besides the loom (rRNA), its building blocks (amino acids on RNA or aa-tRNA), and its instructions (mRNA), other RNAs play more specialized roles in translation. Riboswitches are (mainly bacterial) RNA devices that change mRNA presence or activity in response to molecules in the environment. Using these, an mRNA can not only encode a protein but also express it conditionally, depending on molecular signals in the environment. Transfer messenger RNAs (tmRNAs) are RNAs of hybrid function, as the name suggests, that rescue stalled ribosomes, unable to complete a protein. A core molecule of the membrane mooring for the ribosome (the signal recognition particle) where a protein is being slid through the boundary bilayer membrane is also an RNA, called 7S RNA in bacteria.

RNAs Act to Modify Other RNAs

RNAs act in many capacities to reshape the chemical structures and sequences of other RNAs. Their varied and intricate abilities in this regard certainly suggest that a complicated information economy could be managed using RNAs alone.

In the pathway(s) called RNA processing, RNAs shape new RNAs more recently spun off (transcribed from) their template DNA. The protein enzyme that spins RNAs on a DNA template is called RNA polymerase. In eucaryotes like us, RNAs generally arise as transcripts complementary to one strand of chromosomal DNA's two strands, as a spot of RNA in the cell's nucleus, near the chromosomal location of the DNA information. However, there is a long, multistep refinement pathway ahead for the freshly spun-off RNA.

An RNA intended for ultimate use as mRNA, for example, must be capped (a special RNA nucleotide is added to its 5′ end) and tailed (a string of AAAAs is added to the 3′ end). Most striking of all, sections in the middle, called introns, are cut out (they are partly used but mostly thrown away), and the flanking ends are joined precisely. This process is called splicing, and its orchestration is also in the hands of small RNAs, decorated with numerous proteins; the consequent particle is termed the spliceosome. The RNAs of the spliceosome are small nuclear RNAs (snRNAs). Incidentally, bacteria and cell compartments like chloroplasts also splice, but they use more compact components, called group I and group II RNAs, within the introns. These intron RNA regions carry out self-splicing reactions themselves, without using the spliceosome.

Ribosomes and tRNAs in particular have many modified nucleotides, each of which is one of the standard four ribonucleotides A, C, G, and U, with a few atoms added later during processing. Addition of these modifications is often directed by a protein enzyme, guided by the base paring of an RNA held by the protein. Because modification of rRNA occurs in the nucleolus, a neighborhood within the nucleus, the RNAs are called small nucleolar RNAs (snoRNAs). A few organisms, like the human parasitic protozoans called trypanosomes, wield the same small RNA guide principle to spectacular effect in their subcellular mitochondrial compartment. Trypanosomal mitochondria use so-called guide RNAs to base pair with and correct transcripts that are made without some or most of their Us—they put in C, G, A, and some Us and insert (and delete) Us later! This elaborate RNA processing is called RNA editing, and it is, in its extensive form, limited to this one organism's mitochondria.

Finally, among RNAs that process or modify other RNAs, a likely ancient RNA is part of the enzyme that cuts transcripts destined to be tRNA, thereby creating one end (the 5′) of the final tRNA. This enzyme is called RNAse P, and it is a ribonucleoprotein (an RNA plus proteins) in which the RNA is almost always the catalytic piece—the one that does the cutting.

RNAs Control Genetic Activity

RNA effects double back after transcription to turn up and turn down the activity of the genome at several levels. These effects are built into mRNA messages and also entrusted to different, specialized RNAs that are in charge of the volume of gene expression and nothing else.

Cells have general ways of specifying the use and survival of functional mRNAs, either singly or in groups—these mechanisms, relying on small interfering RNAs (siRNAs) or microRNAs (miRNAs) have only been discovered relatively recently and are not yet completely understood. However, both begin with a double-stranded RNA, which is cut into smaller pieces of 21–30 nucleotide pairs. These short helices are separated, and one strand is usually seized and used to guide a dedicated protein complex to other RNA sequences, perhaps different related sequences, at which it forms base pairs. For siRNAs, which may make perfect base pairs with a target mRNA, one result is that the associated proteins cut the mRNA target, reducing expression of the message. For miRNAs, which may pair partially and with many partially matched mRNA sites, one result is inhibition of translation of the paired mRNA, again reducing expression. The importance of these ways of directing gene activity during the development of complex creatures is still under active discussion, though it is already clear that many essential decisions within the cells of vertebrates like us are enforced by small RNAs.

Instead of acting directly on the transcripts from genes, siRNA- or miRNA-like double-stranded RNA agents may act at an earlier stage of expression, on the structure of the chromatin, and on its accessibility for transcription into RNA. For example, they may prop open genes that have recently been turned on, or they may enforce the inactivation of certain regions of the chromosome that are never expressed, such as the regions used to hitch chromosomes to the division machinery, called the centromeres.

One such example lies at the crux of many male-female differences in mammals. Xist RNA is a long transcript from the human X sex chromosome. Xist RNA directs the condensation of the second X in human females and the subsequent shutoff of its genes. An X chromosome coated with Xist RNA is uniquely chemically modified. This X-inactivation mechanism enforces dosage compensation, ensuring that XX females and XY males get the same dosage of products from their X-linked genes. Unbalanced expression from a whole chromosome is a serious, often lethal, genetic defect.

RNAs Initiate DNA Replication

The regally stable genetic text in DNA cannot replicate without the help of RNA. Perhaps because the DNA assembly enzymes (DNA polymerases) must be so precise, they never work in the loosey-goosey step that hooks the first two freely flopping nucleotides together. That is, DNA polymerases cannot start a new molecule. So to make new DNA when new chromosomes are needed for cell division, pieces of new DNA are initiated by RNA polymerase (sometimes called primase in this role), which can make new chains. These short, base-paired RNA "primers" can be extended by DNA polymerases, but the initial RNA primers are stripped out and replaced by DNA, and the consequent nicks are resealed later.

The list of RNAs and their functions offered so far in this chapter illustrates the ubiquitous influence of RNA within modern cells. DNA is brilliantly suited to long-term storage, sufficiently stable, for example, to survive in ancient bones with little chemical change after millennia. But this DNA-based genetic text cannot be made meaningful without a large apparatus which depends in notable part on RNAs. RNA enables the information in DNA to be expressed by conversion into RNA molecules. RNA enables the replication of DNA so it can speak to the ages. Perhaps most strikingly, the devices that translate genetic structural information into proteins consist mostly of RNA. In particular, the structure that makes the chemical bonds between amino acids in a protein is rRNA. Thus, the ribosome's fundamental reaction is an

RNA reaction, just as it apparently was in the RNA world (see Chapter 17).

The Dogma

Remarkably, there is a standard biologist's prescription that summarizes the several pages of text you have just waded through. This shorter scheme is called the central dogma, a jokingly ecclesiastical reference by Frances Crick to a summary of bulk cellular information flow. On the left in Figure 9.1, DNA serves as the stable store of the cell's information. DNA contains the information for its own replication in its two complementary strands, each of which is a template for the production of the other. DNA can also be transcribed into RNA, and RNA can be reverse transcribed as a DNA, as happens, for example, in the life cycle of an RNA retrovirus like the human immunodeficiency virus. The right-hand arrow is translation (protein synthesis), and it is unidirectional—RNAs that bear the sequence of nucleotides from a gene for a protein are translated into the linear sequence of amino acids that makes up a protein enzyme or a structural protein. The protein then folds spontaneously or with the help of folding catalysts (chaperones) into a protein structural block or chemical catalyst (enzyme).

Figure 9.2 (which combines bacterial and eucaryotic functions) acknowledges several particular functions for RNA described in this chapter's early paragraphs. These tune the expression of the DNA information at all levels of dogmatic genetic activity. RNA directs the modification of DNA itself, to change its transcription (leftmost light gray arrow). An extreme example of such RNA-instructed modification of DNA occurs when a snoRNA templates the synthesis of the end cap

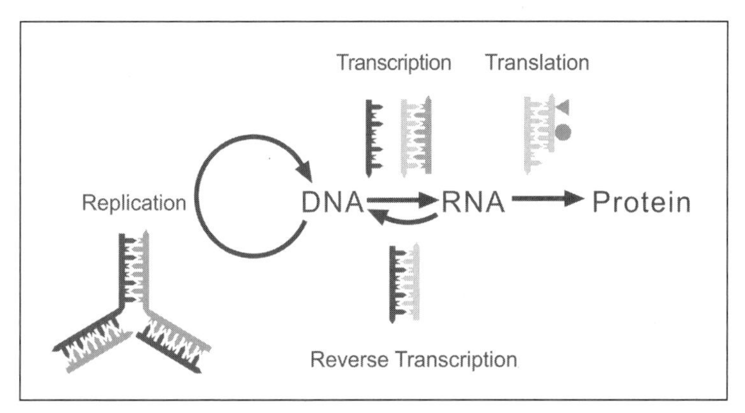

Figure 9.1. The central dogma. Black arrows denote the flow of information, not the conversion of material. Associated pictures of DNA and RNA (thin rectangular backbones with pointy bases) show how nucleic acid base pairing is used to carry out the named processes. Thus DNA specifies its own sequence when it replicates (circular arrow; illustrative dark and medium gray complementary strands base-paired at the left). An ancestral helix (longer dark and medium gray paired strands at bottom left) gives rise to two replica helices below, each with one shorter, newly made strand.

DNA information also specifies the sequence of an RNA transcript (middle). One strand of DNA (medium gray, upper center) is copied to give a light gray RNA transcript.

One strand of DNA (dark gray, lower center) can be made from a light gray template of RNA by reverse transcription, as in the early steps of retrovirus infection of a human cell.

An mRNA (light gray, right) templates the order of (specifies the information in) amino acids (gray shapes) being built into a protein by using base pairing to the amino acids' light gray RNA triplet (3 adjacent nucleotide) anticodons. The RNA triplets with attached gray shapes abbreviate a larger aminoacyl-tRNA molecule that contains the triplet.

of a chromosome (a telomere; horizontal leftward dark gray arrow). RNA primers are the first, temporary, nucleotides during the synthesis of newly replicated pieces of DNA. RNA directs chemical modification of the proteins adhering to chromosomal DNA to modulate such chromatin for different levels of transcriptional performance (e.g., Xist RNA, which shuts off one sex chromosome in XX mammalian females). RNA tunes the performance of complete RNA messages (mRNAs) by depressing their translation (miRNA) or by mak-

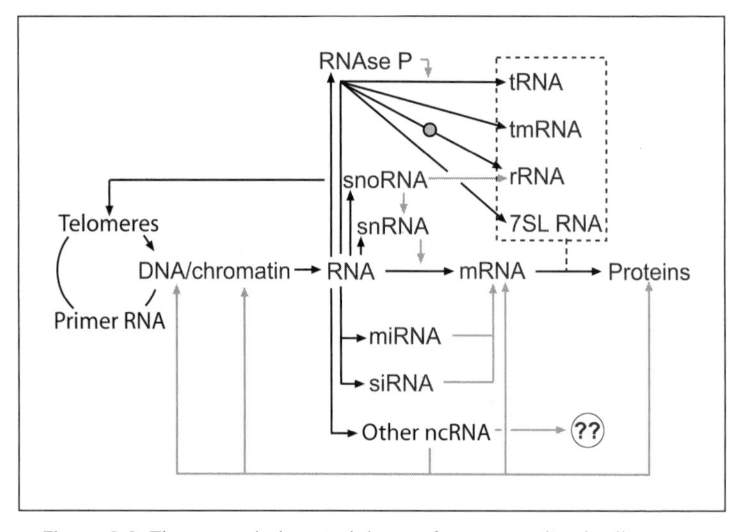

Figure 9.2. The expanded central dogma for a generalized cell. Dark gray arrows represent chemical conversion of the molecule or alternatively informational templating; light gray arrows represent modification of the chemical or higher-order structure (roughly, the shape). The small gray circle shows the spot at which the Cech self-splicing RNA would alter its own structure. The dashed box at the upper right encloses functions of the translational apparatus, which uses RNA information from genes to make proteins.

ing them unstable, directing their destruction (siRNA). snRNAs help to choose sites of mRNA splicing and carry out the deletion and sealing of spliced messages. RNA also directs chemical modification (snoRNA) of other RNAs (e.g., rRNA), apparently to improve ribosomal function. And beyond all these RNA influences on a gene's expression, RNA can bind directly to protein gene products (rightmost light gray arrow) to ratchet their cellular actions up or down.

The horizontal light gray arrow ending in question marks represents the extensive transcription of genes and nongene sections of complex genomes to produce RNAs whose function is yet unknown. Such RNAs embody far more DNA sequences than the sum of all intelligible (gene, protein sequence) transcription already known in complex genomes, notably human genomes. However, it is also possible to think of some of these many novel noncoding RNAs as nonfunctional—due to the imprecision of eucaryotic transcription. For example, the ends of genes are overrun during transcription, and the starts of protein genes influence the synthesis of RNA from other nearby starts in the same chromosomal region. Some of these now-mysterious RNAs, no doubt, are junk—for example, viral sequences that have crashed into a complex genome too recently to have acquired a use.

Some such sequences, however, are conserved in structure across species. Conservation for long periods probably implies that the structure has a use, and these sequences have therefore probably been absorbed into eucaryotic gene regulatory networks. Distinguishing junk from the newly functional RNA sequences that make up a new group of qualitative modifiers is a priority for the next few years, and discoveries in this area will likely enhance the story of RNA functions that we are telling.

Biologists are presently in somewhat the same position as astronomers, who look out at twirling galaxies whose rotation suggests that they have much more mass than can be seen. The physical universe is therefore mostly invisible "dark matter." Surprisingly, the genomes of complex organisms similarly make much more RNA than we can account for. Transcribed DNA is predominantly "dark DNA" whose purpose cannot yet be named. In fact, cosmologists should be more cheerful than biologists; we are mystified by fraction of the sequences in the genomes of complex cells that is far larger than that of the dark fraction of galactic matter. From these dark DNAs come many dark RNAs whose purpose lies beyond the roles listed here. Thus, while there is much RNA that does something specific—something we can point to in a chapter like this —many more RNAs suggest roles not yet known. The question-marked light gray branch in Figure 9.2 is a placeholder for these still nameless RNA functions.

The Message

So what does this add up to? This chapter may seem to have presented an impolite amount of detail. I apologize for resurrecting, compressed into a single chapter, a biology course you may have already taken. But in this case it is really the details that make the point—and it is a worthwhile, nontrivial one. We see the deservedly famous DNA at the left of Figure 9.1—the ultimate instruction that informs the organism. But around the DNA in all directions is a cloud of RNA executors—carrying the information in the DNA into action, organizing the DNA itself within the chromatin, even initiating the replication of the DNA. And even a figure as intricate as this, embodying the present state of our knowledge, seems likely to be critically RNA-incomplete.

DNA is the ultimate source in complex creatures, but its information always flows into RNA structures for use. In fact, because nearly all the information in the DNA genome is embodied in an RNA copy, it is remarkably—and startlingly—easy to imagine DNA fading away without consequence, or to imagine it being replaced by pieces of its halo of RNA, resulting in an informationally equivalent, RNA-only creature. Even today, we are still RNA organisms in this very straightforward sense. Without RNA, a cell would be all archive and no action. If this aphorism seems artificial, consider that if we allowed the RNA to fade away instead of the DNA, nothing—nothing, but *nothing*—of biological importance could happen. We will return to this topic later in Chapter 19, when we talk about the ribocyte.

Thus it may be more realistic to approximate the organism, the organ, or the eucaryotic cell by listing and mapping its RNAs rather than only its DNA. Such a notion might seem to be merely a poetic transformation, but it is instead practical and simple. An organism is not a linear script in a DNA language we have learned to read. In fact, such a simplification is a shocking distance from the truth. Not only does the protein information constitute only 1.2% of the human script, but it is not only linear. In fact it is useless until it is performed at the proper time and placed within three dimensions. Furthermore, the remaining majority of the genetic information is also useful only when drawn out correctly into four dimensions, with attached addresses and times. Looking at Figures 9.1 and 9.2, we see that RNAs are much closer to the four-dimensional plan of the organism than is the DNA script. We need the shapes and localities of varied RNAs before we can understand the development of intricate creatures like ourselves, who sail a complex course through time to adulthood. In the end, the oldest kind of biological information, RNA, may still be closest to the action in the most finely wrought creatures.

Readings

"The expanding world of small RNAs." Helge Grosshans and Witold Filipowicz. *Nature* 451: 414–416 (2008).
An up-to-date review of some small RNA products from complex genomes.

"The expanding transcriptome: The genome as the 'Book of Sand.'" Luis M. Mendes Soares and Juan Valcarcel. *EMBO Journal* 25: 923–931 (2006).
The transcriptome is the sum of all RNAs transcribed from a genome. Here is a comment on the expansion of RNA's role from that embodied in the central dogma to that made necessary by recent discoveries in RNA biology, described in this chapter.

10

RNA Structure: A Tape with a Shape

Nothing exists except atoms and empty space; everything else is opinion.

—Democritus, fifth century BC

We must now build a picture of RNA architecture, as the essential overture to a consideration of where RNA came from and the unexpected variety of things it can do.

First of all, RNA, like all large biomolecules, is a string of small, similar building blocks or subunits. For RNA, these building blocks are nucleotides. "Nucleotide" is chemical shorthand for a molecule with the structure phosphate-sugar-base. Specifically for RNA, it is shorthand for phosphate-ribose-(G, C, A, or U). For example, the string CAUG is a piece of RNA that contains the four usual types of RNA nucleotides. Such an ordered string is called a sequence or a primary structure (panels a and b of Figure 10.1).

Panel b looks at a primary RNA sequence in greater chemical detail. RNA here appears at a higher resolution that shows atoms and the bonds between them in standard chemical argot. We again meet the phosphate–ribose–organic base triune that composes an individual nucleotide. The nucleobases C, A, U, and G (dark shading) are at the top of panel b, the ribose sugar (intermediate shading) is in the middle, and the nucleotide-linking phosphates (no shading) run along the bottom. Phosphates in three dimensions are not squares, but triangular pyramids (tetrahedrons), with the phosphate buried where the pharaoh would be and an oxygen at each vertex. Between each pair of phosphates is a ribose, a carbohydrate sugar (one consisting of carbon, hydrogen, and oxygen) with a 5-membered ring (4 carbons and 1 oxygen) and a linked extra carbon that carries the phosphate. In panel b, at the op-

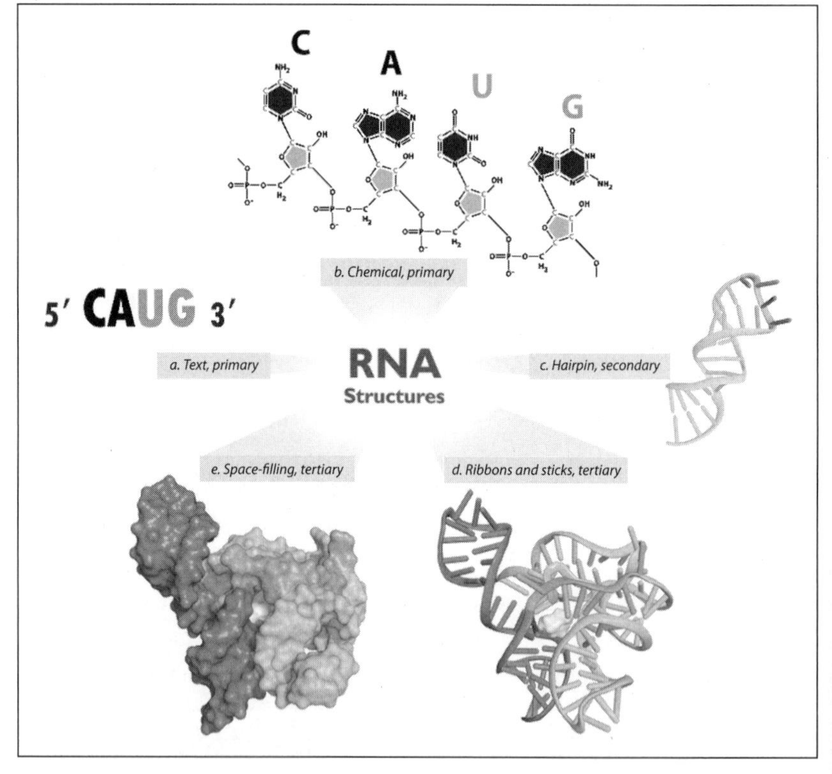

Figure 10.1. Common ways of representing RNA structure.

posite end of each ribose from its phosphate contact with the molecular backbone is one of the four major ribonucleobases; C, A, U, or G.

Chatter about ribose ends emphasizes that nucleotides are asymmetric objects, like an arrowhead, ▶. Thus when you join them to get an RNA chain, you get a chain that has two different ends: ▶▶▶▶ (here shown as flat and pointy). The ends are officially referred to as the 5′ (conventionally on the left) and 3′ (on the right) ends, marked 5′ and 3′ in panel a. In panel b you can see that distinctive naming makes sense because different groups of atoms protrude at the 5′ and 3′ ends. This molecular detail is important because it implies that an RNA backbone has a unique direction—for example, biological RNAs are always assembled from one end (5′, flat) toward the other (3′, pointy).

Among nucleotides' relevant properties is the possession of multiatom areas that act like atomic Velcro, through which suitable collections of atoms stick together. For example, there can be complementary areas of electron deficiency (positive charges) and electron excesses (negative charges). Frequent among these are hydrogen bonds (described in the next paragraph); they, along with other attractive forces, make the RNA chain sticky, both to itself and to water. The result is that a long RNA ribbon does not loll around in water like an anaesthetized snake; instead it readily adheres to water and also folds back on and adheres to itself to make a large folded structure, as a consequence of the interaction of complementary Velcro patches.

More explicitly, oxygens (Os) and nitrogens (Ns) are electron-hungry and filch electrons from coupled atoms to make small positive and negative centers that can attract and repel. In particular, these negative atoms can accept a hydrogen (H) between themselves to form a hydrogen bond. In particular, -OH---N≤ and >NH---O= hydrogen bonds (the dashed lines here) are among the most common in biomolecules. We can see the makings of these H-bond atomic adhesions in the phosphates, ribose, and bases alike in panel b.

One symptom of this tendency for nucleotides to stick to other molecules is that RNA nucleotides attract themselves to form pairs. A associates with U and G with C, using hydrogen bonds between base rings. It is unlikely that anyone now living on the developed Earth has not seen a drawing of the famous DNA helix, in which two backbones going in different directions are bound together by central base-pairing steps to form an iconic spiral. Such multicolored double helices now permeate popular media, selling everything from pills to cars to diapers.

RNA nucleotides also pair and stack to yield helices. In panel c the sugar-phosphate-sugar-phosphate backbone (ribbon) runs helically up one side, loops over (darker section), then helically pairs (using its base sticks) down the other side (in the other direction) to form a hairpin, that is, a base-paired helix (lighter) topped by a central loop (darker). RNA helices with base pairs (panel c) look somewhat like the celebrated (but younger, regrettably immature) DNA helix. RNA base pairs are tilted rather than flattened and stairlike as in the usual DNA helix, but the helical relation is still obvious. When primary RNA strings are bent and looped to form a specific structure (like the helix piece flattened for inspection in Figure 10.2, or like panel c), we say they form a secondary structure, in which the four vertical pairs of nucleotides are the steps of the unseen helical gyre.

Secondary structural elements in turn can combine to form a tertiary structure. As a simple example, hairpin tails can fold back to pair with hairpin loops, creating the intriguingly named pseudoknot (Figure 10.3), in which two helical secondary structures develop.

In this book, we are usually interested in more complex tertiary structures that perform more complicated activities. For example, in panel d of Figure 10.1, several RNA helices and loops in one continuous RNA sequence fold their separate secondary domains together to create a central pocket for a small biochemical, the metabolic molecule *S*-adenosylmethionine (SAM, shown as a light-colored surface among the ribbons and loops). (The separate RNA domains are arbitrarily rendered in different shades so that their assembly can be better appreciated.)

The RNA structure in panel d acts in its native context to detect and communicate the presence of the small central molecule. This RNA fold thereby controls the activity of a gene, which activity is triggered by SAM; it is called a SAM riboswitch. Here is a real tertiary fold formed by sections of the nucleotide sequence looping back reproducibly over each other. In this particular case substructures called helices, hairpins, and pseudoknots (which we have mentioned), as well as kink-turns and three-helix junctions (which we have *not* mentioned) are used to create a pocket for the central metabolite SAM, so that an attached RNA can react usefully to the cell's

Figure 10.2. A short RNA helix or stem.

supply of SAM. Those who wish to know more about RNA folding can consult the readings at the end of the chapter, which make clear that many other recognizable small motifs exist for composing larger active structures like this one. It is also clear that the list of local RNA structural building blocks is not yet complete, even though RNAs with much more complex three-dimensional shapes are already known, like the small rRNA, which has more than 1,500 nucleotides.

Finally, panel e in Figure 10.1 is the same RNA fold as in panel d, with sequences rendered in the same shades, but instead of ribboned backbone and stick bases, it traces the actual surfaces of the atoms, approximating the electronic shape you would actually feel if you had exceedingly tiny fingers. In this so-called space-filling drawing, you can appreciate the size of nucleotides relative to atoms, and you can also see that the RNA tape is rather irregular. Even though a molecule with any substantial number of joined nucleotides is long and thin, a folded RNA still has a complicated shape. It is not locally thin

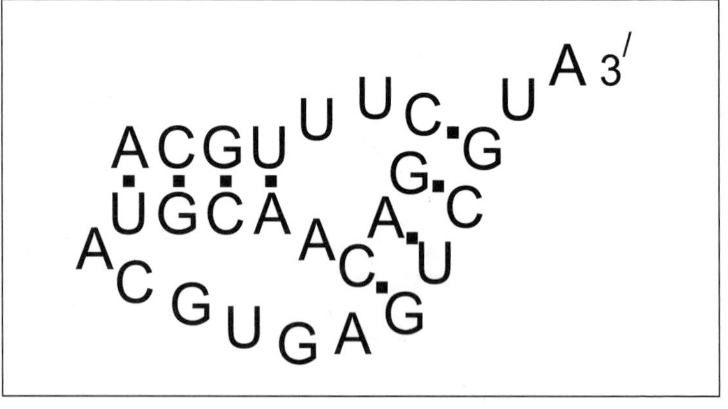

Figure 10.3. Combined helices make an RNA pseudoknot.

and featureless, as suggested by the tape metaphor. Panel e shows most clearly that the large RNA fold makes a tightly fitted pocket for the smaller central molecule (visible through the central open well), almost surrounding it in space. (The origin of the structure used in panels d and e is in the reading by Montange and Batey listed at the end of the chapter.)

RNA structural elaboration can continue into quaternary structure (by putting together two or more of the tertiary structures shown in panels d and e), but we will not follow these higher assemblies in detail. The essential point here is the possibility of large, elaborate structures with specific cavities, formed using only a few elementary RNA interactions. (See the sidebar beginning on the next page.)

These are crucial considerations, because it is the ability of RNA to fold into such a complex three-dimensional crumple, time and again, that makes it of evolutionary importance. RNA usually has a specific, predetermined sequence of nucleotides or Velcro hook-and-loop sections along a strand. RNA structures therefore fold to give dependable holes and furrows, lined by particular available nucleotide atoms. For this reason, RNAs can be built to house other specific molecules on their surfaces, as exemplified by the small SAM molecule in panels d and e of Figure 10.1. To the list of dependably bound items we may add other RNA sequences within the same molecule (to form folded secondary and tertiary structures like those in the figure), as well as other small biomolecules. RNAs can therefore even speed the chemical reactions of the bound molecules. So RNAs can also act as enzymes, sometimes called ribozymes, or enzymes made of RNA (see Chapter 11).

We have seen that RNA is a molecule capable of interacting with itself to give reproducible, elaborated shapes that also stick to (are soluble in) water and are therefore capable of complicated cellular functions. These intricate space-filling structures and their varied and accurate chemical functions—many unknown until recently—are the executors that make an RNA world possible.

RNA structures are made with ease

Here is an easy exercise that illustrates something distinctly important about RNA structure and its particular role in the succession of life on Earth.

On a piece of paper, write 5 different arbitrary sequences for RNAs 25 nucleotides long. Strings of roughly equal numbers of A, U, G, and C of this length are short enough to be within the range of very simple polymerization methods, and molecules of such lengths may well have existed prior to an RNA world.

Insofar as you can, match up A=U and G≡C (Watson-Crick pairs) and G=U (so-called wobble "mispairs"), perhaps by cutting out the sequences you wrote and folding them. Adjacent pairs between sequences running in opposite directions define helical sections (see the figure opposite). You can allow looped nucleotides (because they occur in real RNAs) when they are required between paired nucleotide bases.

Now count the number of paired nucleotides in your 5 RNAs of arbitrary sequence and calculate the percent paired, that is: (paired nucleotides)/25 × 100 = percent paired.

In a typical set of sequences (where, for example, no nucleotide has been used to the exclusion of another), you will find an average of 40–50% pairing even though the original sequences were written with no thought of structure whatever. Typically, you end up with 5 rather complex and different RNA origamis composed of helices and

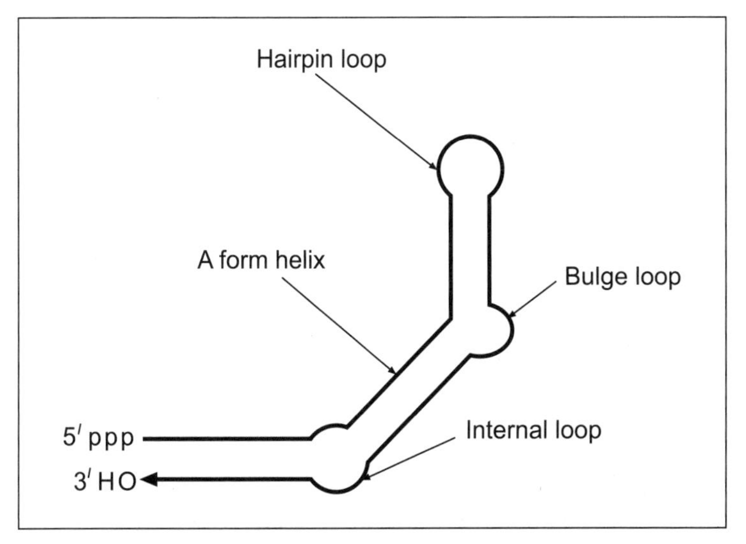

A folded RNA secondary structure flattened for examination (the so-called roadkill representation), with the line representing the ribose sugar-phosphate backbone. RNA helices in this schematic form appear as parallel lines; less regular RNA loops appear as arcs of different kinds.

loops, even though you started with rather short sequences (messenger RNAs, for example, are typically more than a thousand nucleotides long).

The point of the exercise is that, because there are only 4 types of nucleotides and a simple kind of interaction (base pairing), any particular nucleotide has a high probability of finding a complementary one to pair with so it can participate in a small helix. Thus even short arbitrary sequences are about half paired and form varied structures. So, although under primitive conditions you might form only short polymers, relatively complex structures (having varied chemical talents) are still possible. This is particularly so because the helices are not the only active parts; the loops you made contain free nucleotides poised to do potentially useful things (like interacting with other molecules).

Thus though the proteins, composed of 20 different kinds of (amino acid) building blocks, ultimately were better catalysts (and so in time replaced RNA-world RNAs for many cellular purposes), it is possible to see why it might have been preferable to begin with cruder structures composed of a simpler set of 4 RNA nucleotide blocks, which more easily fold into small, but potentially chemically active, structures.

Readings

"Schematic diagrams of secondary and tertiary structure elements." Mark Burkhard, Douglas Turner, and Ignacio Tinoco, Jr. In *The RNA World*, Second Edition, pp. 681–685. Cold Spring Harbor Laboratory Press, Cold Spring Harbor, N.Y. (1999).
A now slightly out-of-date summary of folded RNA structures, but a good way of getting started in thinking about the ways that RNA interacts with itself.

"RNA structural motifs: Building blocks of a modular biomolecule." Donna K. Hendrix, Steven E. Brenner, and Stephen R. Holbrook. *Quarterly Reviews of Biophysics* 38: 221–243 (2006).
A more recent, more inclusive account of RNA substructures.

"Structure of the *S*-adenosylmethionine riboswitch regulatory mRNA element." Rebecca K. Montange and Robert T. Batey. *Nature* 441: 1172–1175 (2006).
This is the original paper about the structure that serves as an example of complex RNA tertiary shapes in this chapter.

11

Intimations of an RNA World

Science is the belief in the ignorance of experts.

—Richard Feynman

Why would one ever suspect that an RNA world and RNA creatures existed? The lost RNA world was a durable theory for 15 years before any experiment strongly supported it. The reason for the persistence of the initial idea lies in the logic of a search for simpler forms of earthly life.

First, recall the delicacy of evolution, embodied in the idea of continuity. Continuity maintains that because successful evolutionary changes are likely to be small, evolved forms must invariably flow via successive small changes from pre-existing forms. This must be as true for triumphant advances as it is for the barely audible, timeless scurrying of multitudes. True late and true early. Continuity should be a reliable guide in even the largest questions—for example, what sort of creature preceded us?

Given that our manner of living must have issued by one feasible step from some simpler creature, our next-nearest ancestor would necessarily be complex (because we are complex) because of the need for continuity. But which among our capacities could we imagine giving up and still be fully capable creatures, able to evolve our full future bag of genetic tricks?

Based on our earlier discussion, we expect a genetic mechanism embodying a viable balance of stability and mutation, to allow for descent with modification. On the other hand, it is hard to see how to do this without catalysts to carry out replication chemistry at a rate useful to cells. Living cells have complicated parts that decay continuously, mandating a limited lifetime. Cells therefore cannot wait indefinitely for complex events to reproduce them. But nucleic acids carry information and do not currently replicate themselves. While proteins direct and accelerate chemistry, they cannot replicate in any known, practicable, direct way.

So, in all modern Earth creatures, genome-encoded proteins called replicases or polymerases are required to make more nucleic acid. However, to make proteins (chains with a specified sequence of amino acids), one needs a complex set of nucleic acids in the ribosome and all its attendant molecules (see Chapter 9). Thus one seems to need a chicken to have eggs as well as the eggs to produce more chickens. The problem of deciding whether chicken or egg could be sacrificed, to leave a simpler evolving system that could progress to the full set of avian capabilities, is particularly acute. The dramatic answer that has emerged is that we can imagine doing without both proteins and DNA.

Soon after the genetic code was established experimentally, circa 1961, thought began about the origins of coding and the translation apparatus. The first serious published effort is due to Carl Woese of the University of Illinois (where he is still in the Microbiology Department). Woese's pioneering work on the triple-domain structure of the Big Tree has already been mentioned. But before and since that research he has thought about the origins of protein biosynthesis. In his 1967 book *The Genetic Code: The Molecular Basis for Gene Expression,* he asks how we might simplify translation, as it currently takes place, to obtain a plausible ancestor for the system that we know.

He answers this continuity question by saying that we cannot imagine that complicated protein enzymes like the aminoacyl-tRNA synthetases (which make aminoacyl-tRNA) could precede sophisticated translation. Nor would it be easy to imagine anything like the complex architecture of the ribonucleoprotein (RNA-protein) ribosome in a primitive system. Instead, the relatively simple molecule that one could imagine early on is tRNA. It is small, and its structure, though sophisticated, is just that of RNA—which even now occurs as a close companion to every growing protein, tRNA being bound by a strong chemical bond to every protein throughout its synthesis.

But how could you make peptides (strings of amino acids)? Forty years ago Woese, imagining a speculative but definite possibility, supposed that primordial tRNA-like molecules had two cavities or sites for amino acids (activated amino acids, that is, amino acids made reactive by chemical modification). tRNA could then bring the reactive amino acids together on its surface and make peptide bonds to join them. Perhaps it could do this more than once, shuffling the previously bound amino acids to one side repetitively, to make short peptides. The need to reuse the same tRNA sites might make the peptides somewhat monotonous (or ambiguous), but this RNA-bound process would be a start that would give translational improvements a place to happen.

We will return to somewhat similar ideas, but the essential notion is that Woese placed the origin of translation within an all-RNA context. Though the RNA world was a concept far in the future, this kind of simplification of early translation was independently plausible to others.

Shortly thereafter, coordinated 1968 reviews from Frances Crick and Leslie Orgel in the *Journal of Molecular Biology* took up overlapping themes. Crick, like Woese, was struck by the complex structure of tRNA, the only RNA whose structure was known at the time. He captured this newly visible tRNA complexity in the suggestion that "tRNA looks like Nature's attempt to make RNA do the job of a protein." Of course this is a twentieth-century perspective; looking forward from the RNA world, protein is Nature's attempt to make peptides do the work of RNA. Whatever the case, Crick takes rRNA and tRNA as the primitive parts of the translation apparatus, and he explicitly envisions that the early translation machine "had no protein at all and consisted entirely of RNA." Similarly, it can be imagined that "the first enzyme was an RNA molecule with replicase properties." Here we have the essence of an all-RNA biota, one that can replicate its RNA sequences and is prepared to begin the production of proteins, encoded amino acid sequences.

Leslie Orgel's 1968 essay turns on a comparison of the ideas that proteins preceded nucleic acids, or alternatively that nucleic acids existed before proteins could be synthesized. The first evolutionary succession (proteins first) is disfavored because there is no example, or even likely possibility, of a mechanism that will replicate sequences of amino acids. While there was no evidence at that time "that polynucleotides have even limited catalytic activity," Orgel presciently leaves the question open. Were it to be true, nucleic acid–based life could exploit the chemicals in its environment, and many new possibilities would arise. In particular, it seems that Orgel saw a salient possibility for the transition to our present bioinformational style: "it is clear at some early stage in the evolution of life the direct association of amino acids with polynucleotides, which later was to evolve into the genetic code, must have begun." This would provide the required transitional routes out of the earlier nucleic acid world, even though Orgel thought that the initial code-forming interactions might have needed contemporary organic chemicals or minerals.

Eight years later, in 1976, a vitamin biochemist from the University of Delaware named Harold White III made a simple observation that notably expanded the idea of the RNA world. He pointed to the fact that many enzymatic cofactors (small reactive molecules bound to protein enzymes to increase their chemical versatility) either were ribonucleotides, containing adenosine monophosphate (a standard ribonucleotide), or else were close to ribonucleotides, with their phosphate-sugar-base structure. Half of all protein catalysts therefore seemed to rely on chemicals that were not amino acids but nucleotides—arguably relics surviving from a previous generation of enzymes that were probably made of nucleotides:

The enzyme cofactor NAD = AMP–nicotinamide nucleotide
The enzyme cofactor CoA = AMP–phosphopantetheine

The reactive parts of this earlier generation, previously residing at the 5′ end of RNA enzymes, now enhanced the reactivity of their descendants, the protein enzymes. This is an extremely persuasive speculation, because the basic insight (cofactors look like ribonucleotides) is true of so many cofactors. It is important to the RNA world because it suggests a complex metabolism with many biochemical reactions catalyzed by RNAs or, perhaps more powerfully, by RNAs with an attached reactive cofactor. Now our hypothetical RNA creatures not only replicate, devise the genetic code, and make peptides, they can also administer a diverse chemistry, as would be required to support and maintain a living creature that replicates and translates.

It was Walter Gilbert, a Nobel laureate in 1980 for his work on DNA sequencing chemistry, who gave the RNA world its unifying name a decade later. His one-page 1986 note in the journal *Nature,* almost invariably referred to by those interested in the RNA world, is nevertheless a stranger in this role. In fact it does not reach out in the directions taken in the preceding writings of Woese, Crick, and Orgel. Instead it is mostly devoted to the development of a Gilbertian idea: that self-splicing RNA—in the style of Tom Cech's group I self-splicing RNA (see just below)—could have been the molecular agent that split genes into exons (to be retained) and introns (to be spliced out). In any case, a complete creature, subject to Darwinian evolution and capable of metabolism, was now a reasonably complete, named mental construct—awaiting only confirming experimental evidence.

That evidence soon appeared. *Tetrahymena* is a pear-shaped protozoan covered with rows of cilia. It can be found rowing about in Colorado ponds and almost anywhere there is standing fresh water. Its appearance—although perhaps beloved by zoologists—is otherwise redolent of a monster movie. But because it is only 0.002 inch long, mankind is in no immediate danger. *Tetrahymena* was the organism selected by Tom Cech for studies in his newly established lab in Boulder, Colorado, because it has many thousands of genes that are transcribed to make rRNA. This suggested that *Tetrahymena* cells would have sizable supplies of the enzymes required to manufacture rRNA, including the enzyme needed to splice out an intron (intervening sequence) that interrupts the large rRNA transcript. In 1981Cech and Art Zaug set out to study this reaction, hoping to isolate the protein enzyme that did the splicing—an experimental feat that had yet to be accomplished at the time.

Zaug obtained the rRNA with its expendable insert directly from cells and cooked it up in the small plastic tubes that host almost all objects of interest to molecular biologists. There was exhilaration in the lab when, on the very first try, splicing apparently worked on the benchtop much as it does in the cell. When nuclear juice from *Tetrahymena* cells was added to the purified rRNA, the intron popped out. Only one thing seemed amiss: in the negative control, to which no nuclear juice had been added, the small intron had still popped out of the larger rRNA transcript. "Well, Art," said Tom, "this looks very encouraging, except that you must have made some mistake making up the control sample."

Or not. Intensive study showed that there had been no mistake. Surprisingly, the RNA didn't need the aid of a protein, but could split out its intron, while sealing the flanking fragments, using its own purely RNA structures to direct and accelerate the reaction.

As frequently happens, somewhat parallel work was under way elsewhere. Sid Altman and Cecilia Guerrier-Takada at Yale were collaborating with Norm Pace, Terry Marsh, and Kathleen Gardiner in Denver to make hybrid RNAse P catalysts. RNAse P is an unusual enzyme which has both a protein and an RNA piece. It is an RNA-cutting enzyme (a nuclease) that forms a tRNA end; that is, it clips an RNA transcript just upstream of a built-in tRNA as a part of the RNA processing pathway that shapes tRNAs for ultimate service in ribosomal translation. (See Chapter 16 and the lexicon for more on this topic.) This RNAse P cut is a universal reaction, required to build the translation machinery in every cell, so it soon attracted study as a widespread, essential biological machine.

In independent, almost simultaneous work in Connecticut and Colorado, it was found that one could put together RNAse P protein and RNA from two distantly related bacteria, and they would collaborate to carry out the RNAse P cutting reaction that releases a tRNA from its transcript. This is remarkable: the proteins and RNAs that now worked together had not met each other for many thousands of millennia. As a part of this mixing-and-matching, it was necessary to try RNAs on their own. In controls in which the proteins were left out (and an unusual amount of salt was present), both Kathleen Gardiner and Cecilia Guerrier-Takada observed that the reaction continued. The RNAse P enzyme activity was present in the purified RNA part, not the protein part. The protein piece did stimulate the reaction, but the entire reaction, a cut beside a specific nucleotide, was carried out somewhat more slowly by the RNA alone.

Both the Cech and Altman and Pace labs had found, in different ways, that RNAs alone could catalyze reactions. And not just any reactions! Both the *Tetrahymena* intron and RNAse P made or broke the bonds between ribonucleotides. This dissipated the chicken-egg paradox, because it allowed one to think that RNAs might string ribonucleotides together (the reverse of taking them apart). RNAs might make more RNAs with no involvement of proteins. They could be the first replicators.

The biological world thereafter contained "ribozymes," the Cech lab's contraction of *ribo*nucleic acid en*zyme*. Proteins were not the only biological catalysts; within the cellular machine were the ghosts of an earlier existence—catalysts made of another material entirely, RNA. Tom Cech and Sid Altman shared the 1989 Nobel Prize in Chemistry for recasting the origins of nucleic acid biology, and of our biology itself, in this way.

These first-found RNA reactions are not marginal in any sense: RNAse P is very accurate, cleaving one particular bond to leave specific tRNA chemical termini. The *Tetrahymena* rRNA intron not only selects two particular nucleotide-nucleotide linkages, but also breaks them, tosses itself out, and reseals its flanking sequences to each other. RNAse P is still the means of processing the end of a transcript containing tRNA in most organisms today. This means that tRNA-like molecules, which probably have had a role in translation since very early before the LUCA, were probably born into a metabolism in which RNA catalysts, like RNAse P, were major agents. In other words, tRNA and protein biosynthesis appeared in an RNA world. This is consistent with conclusions we will draw later, based on broader arguments.

Now we can tentatively solve the problem of envisioning an earlier form of life that would be simpler but continuous in history with our own. Let's make it RNA-only, a ribocyte (RNA cell) that uses RNA for catalysis as well as for storage of information. But is this plausible? In particular, can we suppose that RNA could have carried out reactions besides that of making and breaking the phosphodiester bonds between the nucleotides of DNA or RNA? Remarkably, we can give quite a good answer to that question, in the next few chapters.

Readings

"The RNA world." Walter Gilbert. *Nature* 319: 618 (1986).
Watch the RNA world drawn together, named, and put in context.

"The RNA moiety of ribonuclease P is the catalytic subunit of the enzyme." C. Guerrier-Takada, K. Gardiner, T. Marsh, N. Pace, and S. Altman. *Cell* 35: 849–857 (1983).

"Self-splicing RNA: Autoexcision and autocyclization of the ribosomal RNA intervening sequence of Tetrahymena." K. Kruger, P. J. Grabowski, A. J. Zaug, J. Sands, D. E. Gottschling, and T. R. Cech. *Cell* 31: 147–157 (1982).
The discoveries that identify RNA (as well as protein) as an enzyme.

"Progress toward understanding the origin of the RNA world." Gerald F. Joyce and Leslie E. Orgel. In *The RNA World,* Third Edition, pp. 23–56. Cold Spring Harbor Laboratory Press, Cold Spring Harbor, N.Y. (2006).
A good discussion, with many openings to the original literature, of the problems to be solved in constructing a realistic account of the RNA organism.

12

The Experimentally Impaired Sciences

That which is not observable does not exist.

—Paul A. M. Dirac

Nature's dice are always loaded.

—Ralph Waldo Emerson

We must now face a difficulty in reconstructing the time-out-of-mind that was the RNA world, about 4 Gya. Our story requires acquaintance with some of the most imposingly remote among extinct beings (recall the deep time span in Chapter 5). We must not only acknowledge this gap but face up to its dire implications for our investigation.

This difficulty lies in our inability to bring experiments to bear. In the 400 years since Francis Bacon recommended the scientific primacy of experimental facts, it has been confirmed many times over that carefully designed experiments are the best route to knowledge of the world. Experiments are indispensable precisely because their answers are often unexpected; the outcome of any truly important experiment cannot be guessed.

Yet there are many interesting and essential pieces of information that lie forever beyond the reach of any experiment—those curious about ancient life on Earth are not isolated in their remoteness from the objects of their curiosity. Where did the moon come from? What did Alexander the Great's speaking voice sound like? Whom should we blame for the hot spell we're currently experiencing? You no doubt detect a common element in these questions: whenever there is a historical dimension to our interests, many questions cannot be answered by experiment. It is not only time that can bar us from our preeminent investigative tools; there may also be insurmountable barriers of distance and scale. Sometimes these barriers combine to form an even more impenetrable obstacle. What killed the dinosaurs? No conceivable experiment can immediately bear on the question, whose objects lie long ago, far away, and on a planetary scale that baffles intuition.

However, such scientific pessimism is not completely appropriate, as you already know. There are strategies that will enable us to reach over barriers of time and magnitude. One such approach relies on Bayes' theorem, a notion developed by the nineteenth-century British cleric Thomas Bayes. Its basic idea is simple: given a reasonably complete set of ideas about the past (or about anything), the most probable history is the one most likely to have yielded what you *can* see. This sounds like avuncular advice from the good-spirited Reverend Tom Bayes, but in its full-bore mathematical form, it is a logical, numerically precise necessity within the foundations of probability calculus. As a result, Bayes' theorem enables us to estimate our most probable likely history by asking a current, potentially accessible question—what past circumstance is most likely to have given rise to what we now see?

For one example, consider the search for the crater from the impact that, according to one theory, killed the dinosaurs. The finding of such a crater at Chicxulub in the Yucatan—sufficiently large and dating to 65 million years ago (the so-called K/T boundary, the point in time at which the dinosaurs vanish)—therefore makes the idea of a lethal effect on the Earth's creatures from impact quite a bit more likely. However, such Bayesian snooping is not an infallible procedure. For example, we must have a list of ideas to test that includes the correct explanation. Otherwise, no comparative test of the probabilities of our list can lead us to the right guess. For the K/T extinctions, it took decades of resolute discussion for the most useful theory to float to the top, so this is not a trivial difficulty. Nevertheless, distinguishing between historical explanations (or evolutionary possibilities) and declaring one of them most probable is frequently a usable strategy.

The other principle that can help us is continuity. That is, the progress of evolution is plausibly linear (with branches when new species arise), like the trees that represent the resulting relations between organisms (see Chapter 3). There is a deep, nontrivial point to calling evolutionary change linear. At present, all new creatures on the current Earth come from creatures who existed before (there was, at least once, an exception at the origin of life, but continuity can help us even there). For example, continuity might suggest that the apparent discontinuity at abiogenesis is one of definition, rather than an intrinsic property of the event (see Chapter 6).

Because a "newly appeared" creature is a modified form of an old one—better fitting, say, a changed environment—it resembles the old one and is genetically continuous with it. Therefore, as Darwin and Darwinism have often emphasized, individual evolutionary changes (mutations that are selected) are hugely likely to be small, and change (viewed on a short time scale) will be virtually continuous. Make no mistake: as emphasized elsewhere in this book, genomes are ultimately digital information, so there is necessarily a tiny jump, a smallest finite change. This size can be submerged in the notion of continuity because it can in fact be a very small effect. In fact, large changes in genetic texts are possible; inversions, deletions, and duplications of patches can alter large tracts of nucleotides. But useful genetic changes can only slightly alter a genome's expression without completely wrecking development and behavior. So even these sizable textual events must have (relatively) subtle overall effects, to allow a complex organism to survive.

Moreover, though one might think otherwise, continuity is not challenged by the evident truth of the view that catastrophic and unpredictable events have shaped Earth's biota. Catastrophes necessarily act only in the context of the variation among species that actually experience the cataclysm. The small mammals that crawled out to inherit the Earth after the Chicxulub K/T impact had wiped away the dinosaur dominion were a selection from preexisting mammals that saw the impact and went into hiding. They could not have been very different, though they might have represented a subset of likely survivors. But only in the thousands of generations following the impact could they substantially evolve to suit a changed world. Change can easily be so radical that no existing variant of a creature can survive (consider the K/T dinosaurs). Therefore, the organism that does in fact survive is necessarily recognizable as a cousin of the one that existed before (consider the K/T mammals). Viewed in foreshortened perspective, with many generations of selection in between, selection can appear to have leaped across a catastrophe. But viewed close up it cannot, and does not, leap.

Bayesian thinking helps us make deductions about descent, because it is present thinking about evidence of past events. Continuity is even more powerful in one sense: when it can be applied, it is two-faced, looking both back and forward in time. A continuous function, in mathematics, is one that converges to a unique value as it approaches a point from the right *or* the left. So continuity carries implications about both the future and the past.

Bayes' theorem and the principle of continuity combine to restrict the course of evolution, at least in theory. Continuity limits evolution to small steps that form a smooth trail through any progenitor; Bayes says that the direction of the trail must be the most probable when viewed in the light of ancestral genomic properties and environment. When, in the future, we formulate regularities in the route(s) taken by evolutionary change, we can hope to compute the likely course of evolutionary change and, for example, recognize when unexpected events (like a plummeting asteroid) must have redirected an outcome.

So to the ribocyte: it is different from a modern organism in its rate of evolution. It probably has an unstable number of genes. Exact division of genomes is complicated, so it is more plausible that the ribocyte is dependent on random distribution of the pieces of its RNA genome when it divides. It probably also has a lower accuracy of replication owing to less refined RNA replicase activity. The apparatus that translates the RNA genome is also segregating and mutating, superposing another layer of variation in the expression of the ribocyte's genetic information. All these things combine to make it quite certain that initially identical cells would be genetically different, after only a few duplications.

This fate contrasts with the constancy of modern cells. No matter how many times a dog's cells replicate, they stay canine. But the dividing ribocyte, in contrast, had immense genetic variance. The rate of evolutionary change is proportionate to the genetic variance (this is the fundamental theorem of natural selection, due to R. A. Fisher). Thus the ribocyte probably lived amid a blaze of evolutionary change compared to more familiar modern organisms. This is useful to remember while generalizing about its properties, which are usually (as here) visualized as constant.

In any case, we can now easily make several Bayesian predictions about the RNA world. The RNA world appears in its definitive form when RNA learns to replicate and, as a result, Darwinian evolution begins. That is to say, we predict that RNA is capable of reactions that replicate itself, and this prediction has been partially confirmed (see Chapter 15).

We can also predict the major event at the fall of the RNA world. The RNA era was succeeded by our nucleoprotein world, in which proteins are the major catalysts. For what may be simple reasons, biological catalysis begins with RNAs (because simpler structures are more chemically proficient; see the sidebar on pages 106–107), but ultimately the greater chemical versatility of the 20 amino acids was decisive. Primordial RNA catalysts, with their 4 chemical kinds of monomers and cofactor groupings, are replaced by more efficient proteins (with 20 or more kinds of monomers *and* salvaged bits of RNA cofactor). Thus the invention of translation, foolishly set loose by the dominant RNAs of the RNA world, brings the proteins into existence and ends the era when RNA did everything. We accordingly predict that RNA must be able to perform all the reactions required for synthesis of coded proteins (translation) on its own. Because RNA does not presently carry out all these reactions, confirmation of these predictions will validate the idea of continuity of our era with an RNA world and will, by Bayesian reasoning, raise the probability that RNA creatures were our immediate, now lost, ancestors.

Reading

"Origins of the genetic code: The escaped triplet theory." Michael Yarus, J. Gregory Caporaso, and Rob Knight. *Annual Reviews of Biochemistry* 74: 179–198 (2005).
The last part of this review discusses Bayesian investigation brought to bear on the study of molecular evolution.

13

Test Tube RNA Evolution: First Light

> There is nothing exempt from the peril of mutation; the earth, heavens, and whole world is thereunto subject.
>
> —Sir Walter Raleigh

RNA was the first molecule to be scrutinized in a laboratory during the act of evolution. Such Darwinian voyeurism began in the laboratory of Sol Spiegelman at the University of Illinois in the late 1960s. We examine these events now as part of our effort to make the existence of RNA organisms probable by exhibiting modern evolution of varied functional RNAs on lab benchtops.

The Spiegelman laboratory had been concerned for some time with the replication of the bacterial RNA viruses Qβ and MS-2. These viruses (called bacteriophages) live by invading bacterial cells and issuing new instructions to make viral particles, killing the bacterial cells in the process. The members of the Spiegelman lab were particularly interested in RNA genome replication, the conversion of incoming single-stranded viral RNA into new copies of viral RNA for assembly into descendant viral particles.

They made significant progress on viral RNA replication, ending with the purification of the multipiece (protein) replicase enzyme. When that purified replicase was given a viral RNA template to instruct it and ribonucleotides to polymerize in a simple salt solution, it would make more viral RNA without any other requirements. The efficient reproduction of the viral genome in reaction tubes suggested that a "Darwinian experiment" could be carried out in vitro (in glass; though in plastico would be more accurate) to study viral RNA evolution.

The Spiegelman lab achieved this goal by diluting a starting mix of nucleotides and RNA and replicase after replication had occurred, then transferring a sample into a fresh tube with more enzyme and nucleotides, but none of the previous RNA, and repeating the process. In this way the starting RNA from the first tube would replicate in successive pure, ideal environments. After days of transfers to new tubes every 10–20 minutes (with overnight breaks; 75 reactions in all), they could ask: "What would happen to the RNA molecules if the only demand on them is the Biblical injunction, *multiply,* with the proviso that they do so as rapidly as possible?" Given the tone of the question, the answer was almost comic.

After synthesis of about 750-fold the total amount of RNA added to the first tube, the replicating viral RNA was only one-sixth as long as the viral RNA that had first been added. This was established using techniques that directly measure RNA size, but it was evident in other properties that would change when five-sixths of a viral genome was tossed out. For example, intact full-size viral RNA produces complete virus when the RNA is introduced into naked bacterial cells. This viral production of more fully active viral RNA tailed off and did not survive the fifth transfer. Winning the Spiegelman lab's race for replication was quickly and completely fatal to the intact initial viral RNA.

Thus, ironically and without planning, these initial experiments produced among the most impressive RNA evolutions ever observed. Freed from the need to preserve any genomic capacity except the ability to replicate, Qβ viral RNA discarded everything except the end of the RNA where the replicase protein(s) require short nucleotide sequences in order to begin tracking along, assembling an RNA copy. All other information—thousands of nucleotides—became a burden that only slowed the appearance of a replica. Because of the ready availability of accidentally fragmented RNAs in the reactions, complete but slowly replicating RNA genomes lost the race to zippy RNA fragments. The power of selection to shape RNA could hardly be more clearly demonstrated than by this decisive abbreviation.

However, these same techniques could be used to evolve more subtle variants of the starting viral RNA. (See also the sidebar beginning on the facing page.) Using a shortened RNA already replicating at a nearly maximal rate, the Spiegelman group allowed 108 serial reactions in the presence of increasing amounts of a dye that binds to RNA and consequently partially gums up its replication. The result was the selection of RNAs that appeared to retain about the same (reduced) size, but that seemed to bind the dye less well. The newly selected RNAs continued replication in dye concentrations that stopped the initial RNA in its helical tracks. These Qβ RNAs accordingly evolved a weaker interaction with a chemical toxin. This outcome introduces the idea of laboratory evolution of RNAs with new properties—an idea that we will develop in the next chapter.

RNAs that appear from nowhere

Remarkable stories about Qβ viral RNA evolution in the laboratory do not end with the information presented thus far in this chapter. This narrative illustrates both science's ability to collectively create a more accurate explanation and also how new light can be shed on an evolutionary topic by laboratory experiments. Both ideas reappear later in this book.

It has been argued that replicating RNAs not only evolve *in plastico* but also appear spontaneously in solutions that contain nucleotides but no preexisting Qβ RNA. This even more astonishing notion is on exhibit in the work of the German biophysicist Manfred Eigen. Eigen became a Nobel laureate in 1967 for devising ways to study fast (bio)chemical reactions. He was deservedly seen as a heavy hitter, both analytically and experimentally. So it was widely noticed when Eigen and collaborators in Göttingen cooked up high concentrations of pure viral replicase protein and nucleotides, *adding no template RNA at all.* After a long time with nothing happening (unlike the prompt duplication that follows when RNA replicators preexist), small RNAs of 35–50 nucleotides appear, seemingly created spontaneously.

These new-found RNAs are varied in structure and usually poor replicators. As you might now hope, they evolve as they replicate, doing so more and more rapidly, producing sequences better and better adapted to their test tube environment. It seems that ribonucleotides have a sufficient tendency to assemble that high replicase and nucleotide concentrations produce varied replicators, after a lag, with no prior RNA instruction at all. The variability from reaction to reaction, the long lags, and the absence of apparent sequence relation to the original virus led the Eigen laboratory to argue that this was a new reaction; new RNA replicators were created by the replicase. While this is not a serviceable model of abiogenesis (because high concentrations of the complex, purified protein replicase and nucleotides are required), it still seemed a rousing example of latent creative potential in protein-nucleotide solutions.

However, these ideas appeared in a new light when David Hill and Tom Blumenthal at Indiana University observed that the protein replicase, purified in a different way, did not support the appearance of "spontaneous" RNA replicators. Even more telling, when they killed the replicase protein in purified "spontaneous" preparations (by heating) but left potential RNAs intact, "killed" preparations restored spontaneous synthesis when added to other purified replicase made non-spontaneous by purification. In other words, it seemed that there must be something, perhaps an undetected RNA template, that allows "spontaneous" replication even in these highly purified proteins. But how might we explain the variability, lags, and unusual sequences that had originally suggested RNA spontaneity in the Eigen experiments?

The unforeseen answer appeared several years later from Alexander Chetverin and colleagues in Moscow. By using a sensitive new way of detecting replication, they found that small RNA replicators could travel through laboratory air, perhaps in accidental aerosols or on dust particles. Any lab that had once harbored small viral RNA replicators might therefore be infected, and it might infect new solutions brought into the room with a varied and varying population of small RNAs that can duplicate in the later presence of protein replicase and nucleotides.

So the creative potential within viral replicase and nucleotides is probably an illusion, but a population of ribocytes exchanging RNAs through the air becomes conceivable. Aerial genetics has not since become a recognized topic in biology. However, genes jumping through the air between RNA populations seems every bit as astonishing as the initial idea of spontaneously appearing RNA.

Readings

"Template-directed and template-free RNA synthesis by Qβ replicase." C. K. Biebricher, M. Eigen, and J. S. McCaskill, *Journal of Molecular Biology* 231: 175–179 (1993).
A summary review of the evidence concerning the appearance of the small RNA replicators in apparently untemplated reactions.

"On the nature of spontaneous RNA synthesis by Qβ replicase." A. B. Chetverin, H. V. Chetverina, and A. V. Munishkin. *Journal of Molecular Biology* 222: 3–9 (1991).
RNA goes airborne, and lands and colonizes new areas favorable to its replication.

"Does Qβ replicase synthesize RNA in the absence of template?" D. Hill and T. Blumenthal. *Nature* 301: 350–352 (1983).
Spontaneous replication might be cryptic RNA contamination.

"An extracellular Darwinian experiment with a self-duplicating nucleic acid molecule." D. R. Mills, R. L. Peterson, and S. Spiegelman. *Proceedings of the National Academy of Sciences of the USA* 58: 217–224 (1967).
RNA begins its directed laboratory evolution.

"*In vitro* selection of bacteriophage Qβ ribonucleic acid variants resistant to ethidium bromide." R. Saffhill, H. Schneider-Bernloehr, L. E. Orgel, and S. Spiegelman. *Journal of Molecular Biology* 51: 531–539 (1970).
RNA evolves a new response to an RNA-binding "drug."

14

Selection Amplification: Interrogating RNA's Possibilities

> The theory of evolution by cumulative natural selection is the only theory we know of that is in principle capable of explaining the existence of organized complexity.
>
> —Richard Dawkins

Now we can discuss the approximate recreation of RNA activities for an RNA world using benchtop evolution. First I sketch the problem posed by the yawning volumes that exist within the imaginary landscape sometimes called "sequence space," an imaginary kingdom that contains all RNA sequences.

Here is the dilemma: there are four kinds of ribonucleotide (A, U, G, and C) and therefore four kinds of "RNA" one nucleotide long. It begins to be a little more realistic to talk about a short but recognizable RNA two nucleotides long (a dinucleotide). There are $4 \times 4 = 16 = 4^2$ kinds (sequences for) these, any of four nucleotides at the first position with any of four at the second position. In order to make it likely that we have each of the possible sequences, picked blindly from a box, we must have roughly six times the number of total sequences that exist. So if we had $6 \times 16 = 84$ molecules of RNA two nucleotides long in our box, we would be reasonably sure that we had all sixteen possible sequences (actually 99.75% sure). It is clear why we need several times the $4^2 = 16$ possible sequences: we cannot be sure of picking each of the possible molecules if we blindly pick only once per desired sequence. The calculation expands easily: if we want the number of sequences *plunk* long, there are 4^{plunk} to be found for that length:

$4^2 = 4 \times 4 = 16$ sequences two nucleotides long,
$4^3 = 4 \times 4 \times 4 = 64$ sequences three nucleotides long,
$4^4 = 4 \times 4 \times 4 \times 4 = 256$ sequences four nucleotides long,

and so on.

These notions may seem a bit abstract, but they have a point that cuts: 84 copies of a dinucleotide RNA weigh about 9×10^{-20} grams, an invisible speck by anyone's standards. (There are 28.35 grams in an ounce.) But suppose that we wanted to be sure that we had every copy of an RNA sequence 50 nucleotides long; that would be 6×4^{50} sequences or about 4×10^5 metric tons! In short, because the size of the RNA is in the exponent, the amount of total RNA we need to cover bigger and bigger sequences expands impressively as RNA size grows. So the size of the RNA sequence we can possibly find is determined directly by the size of the pile we can search. This means that the size of the sequences that could have been called up for duty back in the day of the ribocyte would have been strictly limited by the amount of RNA that could have plausibly been present.

So let's talk about the problem we're actually interested in: to set a useful limit, we will also make an insanely optimistic calculation. Our particular insanity assumes that RNA was once one of the most abundant chemicals on Earth. For example, suppose that RNA was once as abundant as the salt in the sea, and we could search through that sea of RNA for sequences. What length might those RNA sequences be? It is admittedly mad, or at least hugely eccentric, to think that there ever was a time when RNA made by natural nonbiological processes was as abundant as sea salt, but that's what I mean by "insanely optimistic." The sequences that we could actually find on Earth during the dawn of the RNA world must have been considerably smaller than those we will get from this over-the-top calculation. Afterward, we can set out a more levelheaded view.

There are about 1.4×10^{24} cubic centimeters of sea water on the Earth today. If, for simplicity, we take all the salt in the sea to be table salt (NaCl), there would be a total of about 5×10^{22} grams of sea salt. (See the sidebar on pages 62–63 if you want a review of numbers in this exponential form.) Just to be clear, that's 5×10^{16} metric tons (fifty quadrillion American tons or 50 petatons) of salt, which if poured out as a powder would make a pile 500–600 kilometers across—the size of Kansas. The salt mound would also be the highest mountain on Earth, towering as high as 20 Everests, poking out above the atmosphere.

What size RNA sequence can we be confident we could find in this vast heap, if it were RNA? The somewhat surprising answer is RNAs about 68 nucleotides long, definitively shorter than a tRNA and much, much shorter than the large RNAs inside ribosomes, which run on for thousands of nucleotides. Now a crucial difficulty stares us in the face. We have a severe problem—unavoidable because it is based on obligatory numerical considerations. Neither our avatar, gazing up at the towering salt pile, nor evolution on Earth in all its power could reasonably be expected to find more than short RNA sequences compared to the many that cells use. If these shortmers are not big enough to do the jobs a ribocyte will need (even employing crazy assumptions about RNA abundance), the concept of an RNA world as a period populated by organisms that depend on functional RNAs is dead on arrival. Resolution of this question is the duty of this chapter.

In fact, the stunning outcome is that small RNAs turn out to be unexpectedly capable—more capable than anyone knew or predicted at the outset. Surprisingly small RNAs can do many chemical and physical jobs (see Chapter 10). But to see how this can be shown in a real-world experiment, we need to understand selection-amplification, so that is the next topic.

Selection-amplification or SELEX (SElection of Ligands by EXponential amplification) is a way of asking if a particular population of RNAs can do some new task. We wish, for example, to know if RNA can perform biological reactions that it does not carry out in modern cells. Figure 14.1 shows the trick: because nucleic acids (RNA and DNA) can replicate, we can repetitively use any routine that physically separates a minority of capable nucleic acids (the selection) to purify those capable ones—we just replicate them after selecting them. The difficult part is usually to devise a chemical purification trick that physically pulls out the small initial minority of RNAs that can perform some unknown reaction.

At the left of the figure, we have large numbers of RNAs of arbitrary sequence, often made by synthesizing RNA on a DNA template that has been made with all 4 nucleotides at most transcribed positions. Such randomized DNA templates can be made in a robotic nucleic acid synthesizer, for example. The leftward random-sequence RNAs fold into a similarly

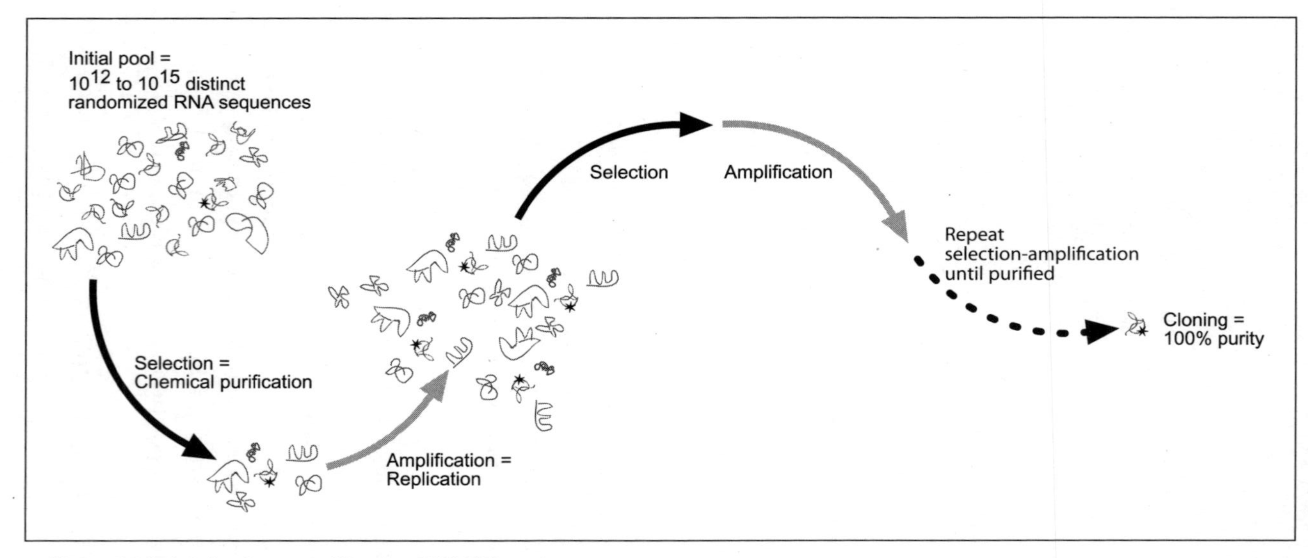

Figure 14.1. A selection-amplification (SELEX) cycle.

varied collection of structures, which the figure roughly shows as squiggles. Among these, a minority (say one in a trillion, occasionally as few as one molecule in the entire population) can perform a new reaction (small star). This tiny fraction of the population would usually be lost to us, unseen and unknown in the great pile of inactive molecules. However, suppose that this interesting and rare minority are partially purified, using their unique chemical qualities as a handle. For example, we might collect the stars by having them stick to a surface with a chosen biochemical immobilized on it; this would select those rare RNAs that fold to form a surface patch that attracts the biochemical.

After selection (bottom left) the RNA population is much smaller, because it includes fewer of the inactive molecules; the active RNAs (with stars) are therefore much more prominent. However, when only one in a trillion starting molecules will carry out the selected reaction, there will be many duds that also survive the first step. This partially purified population is then amplified by various RNA replication strategies. The rightmost RNA population has been replicated—we have more starred molecules, but also more of the inactive ones that sneaked through the selection even though they do not have the activity. This is the inevitable result of the imperfections inherent in any real chemical fractionation. (Compare the discussion of chemical mistakes in Chapter 4.)

We now carry out the selection again, further concentrating the active (starred) RNAs and further rejecting the slackers that were mistakenly retained. Because of the uniquely practical replication of nucleic acids, we can crank in as many selection steps as are required to make the initially minute active minority into a large fraction of the total population. A real experiment might require 5–8 turns through a SELEX cycle.

Finally we use molecular cloning, which puts each RNA sequence into a single matched DNA molecule for DNA sequencing or further transcription. Thus cloning also completes the purification process because it separates the population into individually pure sequences—clones.

The outcome of the overall SELEX experiment is quite remarkable. It literally isolates molecules that might have started out as one among 10^{15}; that is, it potentially makes active molecules 10^{15}-fold (1,000,000,000,000,000; an American quadrillion-fold) more frequent, quite likely yielding a 100% pure product. A quadrillion-fold is the difference between knowing I am on the surface of the Earth somewhere, land or water, and knowing that I am in my office chair in Boulder, typing this paragraph. No other process in science or technology appears to rival this powerful, cyclic, biology-based purification in its ability to resolve and purify a rare substance. SELEX results have consequently transformed the search for new RNAs, to say the least.

We finish this chapter by considering the question of the quantitative threshold for the RNA world. This is an oppositely directed version of the sea salt calculation, in order to emphasize a complementary conclusion: how much RNA must we collect to start an RNA world, at minimum? The trend of the answer seems strange at first, because the number is small. This is because the size of the probably-found RNA increases only slowly as the pile of RNA we search increases. This fact restricted what we could find in a Kansas-size load of salt, but it helps us now. Restated for the new question, the size of the probably-occurring RNA also shrinks only relatively slowly as the haystack of RNA to be searched for this active needle decreases.

Furthermore, the smallest useful pile is evolutionarily crucial —as emphasized several times in this book, RNA is a chemically difficult object for both natural and geochemical synthesis. Assume that enough RNA was needed to allow the occurrence of active folded molecules, so evolution could find them. At this point RNA sites could have been adopted for chemical purposes by organisms of the time. I call this the "axiom of origin" to emphasize that it is not a law but merely a plausible definition for the threshold to the RNA world. Considerations like those that began this chapter—along with an attempt to account for the fact that not all sequences will fold correctly, even if all the required nucleotides are in place—give an answer. The amount of randomized RNA in a few tens of thousands of bacteria or a few tens of human cells is enough. This is tens of nanograms of RNA molecules (10 or 20×10^{-9} g). This is a minute dot, but it would have been enough to find RNA binding sites and simple ribozymes. Furthermore, these calculations have been confirmed by SELEX experiments that find real active RNAs at the predicted frequencies in real RNA populations. As we have seen already in the sidebar on pages 106–107, the statistical limit in the size of the RNAs we can find is more than compensated for because small RNA molecules, with the equivalent of 15–20 working nucleotides, unexpectedly still fold to give many useful activities.

As our next task, we sketch what SELEX suggests about the mandatory elements for the RNA world. Prominent among these are the RNA replicator that probably initiated Darwinian selection, and the translation apparatus that closed the RNA era by elaborating a new generation of superior catalysts, the enzyme proteins.

Readings

"Ribozyme catalysis of metabolism in the RNA world." Xi Chen, Na Li, and Andrew Ellington. *Chemistry and Biodiversity* 4: 633–655 (2007).

An up-to-date guide to recently selected RNAs that could have been useful to a ribocyte.

"*In vitro* selection of RNA molecules that bind specific ligands." A. D. Ellington and J. W. Szostak. *Nature* 346: 818–822 (1990).

"Selection *in vitro* of an RNA enzyme that specifically cleaves DNA." D. L. Robertson and G. F. Joyce. *Nature* 344: 467–468 (1990).

"Systematic evolution of ligands by exponential enrichment: RNA ligands to bacteriophage T4 DNA polymerase. C. Tuerk and L. Gold. *Science* 249: 505–510 (1990).

The founding documents of SELEX, describing three virtually simultaneous, somewhat different selective techniques for discovering new RNA activities in populations of varied sequence.

"Diversity of oligonucleotide functions. Larry Gold, Barry Polisky, Olke Uhlenbeck, and Mike Yarus. *Annual Reviews in Biochemistry* 64: 763–797 (1995).

An early review of the use, utility, and results of benchtop selection applied to RNA.

15

RNA Duplication: Replicase Activity in Real RNAs

So that's the thought: macromolecule, metabolism, replication.

—Cyril Ponnamperuma

Cyril Ponnamperuma's chastening remark reminds us that there are views other than mine, which considers replication and consequent Darwinian evolution to be first, foremost, and indispensable. In fact, the earliest RNA replicase, an RNA that replicates RNAs in its environment (including, perhaps, those like itself) has already appeared several times in these pages, cast as the simplest Darwinian system. In this recurring fable, RNAs that replicate other RNAs must be possible. But is such an RNA-world replicase realistic?

Replication has occupied many people concerned with the RNA world, because it arguably isolates the key issue regarding the existence of the RNA epoch. If the answer to the question of replication is "no," the whole notion of an RNA world founders, or at least must be rebuilt on a different footing. This chapter gives a brief snapshot of experimental attempts to answer the question. As you can anticipate from the previous chapter, we particularly want to know how a selection can be laid out to ensnare RNAs that perform replication-like reactions.

Today RNAs are made by assembling nucleotides, using the chemical energy from disruption of the triphosphate (the 3 linear phosphates) appended to the sugar-base of nucleotides. This means that nucleotides are assembled keeping the innermost of the three phosphates, creating so-called phosphodiester bonds. Thus the replicase question can be stated in the language of chemistry: can RNAs make phosphodiester bonds? We are already entitled to suspect that they can, because the founding natural ribozymes carry out related reactions. RNAse P breaks a phosphodiester bond, inserting H_2O. The self-splicing *Tetrahymena* RNA both makes and breaks them when it cuts itself out and seals the gap thus made. Therefore, the only question remaining may be how to make phosphodiester bond reactions run in a direction that would form new nucleotide chains.

The first attempts to demonstrate that ribozymes carry out reactions like RNA replication (act as RNA replicases) were in fact based on persuading a *Tetrahymena* self-splicing intron to use its diverse preexisting abilities. For example, when the *Tetrahymena* intron joins exons, it is making a phosphodiester bond. Therefore we might envision constructing a situation in which the *Tetrahymena* group I RNA, evolved to perform splicing reactions, is essentially splicing exons over and over, each time extending the first products by a one-nucleotide "exon."

The chapter's reading list points to two such examples, in which the self-splicing intron joins long or short ribonucleotides, using reactions modeled on the forward or reverse steps of RNA self-splicing. Impressively, if you give a close relative of the self-splicing RNA relatively long oligonucleotides already paired next to each other on a complementary strand, it will seal the nicks between pieces to complete a continuous replica of the template strand from the oligonucleotides you supplied.

However, this method requires that the nicked template-partial replicas be available. An RNA that regularly runs across large partial copies of itself lying about in the environment is clearly a special case. We would rather have a general activity that would replicate a template RNA of any sequence. The easiest way to imagine such general replication is to envision 4 activated (reactive) nucleotides that link up in sequence one-by-one under the (base-pairing) instruction of the template.

Such a method would mimic biological reactions most closely if the activated nucleotides were the nucleoside triphosphates universally used by real organisms for nucleic acid synthesis today. (Their universality implies that triphosphates are ancient RNA building blocks.) After following the most successful such experiments, we can affirm that RNA has roughly the required stuff (Figure 15.1) but presently copies only about one to two turns of RNA helix.

The top panel of the figure shows the elements of the desired reaction. The structures are nucleotides, and two (G + U) nucleotides at the left become joined by a phosphodiester bond on the right (in GU RNA). Nucleotides—with (darker) nucleobases pointed upward, (lighter) sideways sugar pentagons, and three phosphates—join by sacrificing (kicking out) their two terminal phosphates. Joined nucleotides retain the internal phosphate (top of the figure) between them in the phosphodiester bond that always chains together RNA and DNA strands. This portion of the figure shows chemical events common to reactions mediated by enzymes called ligases, polymerases, or replicases in biological jargon. In the top panel there is no template strand directing the correct nucleotides to be joined, but that feature is added in the drawing below. There a larger RNA with several chemically active regions (space-filling shapes) facilitates a reaction that adds a nucleotide opposite a base-pairing partner in the template strand

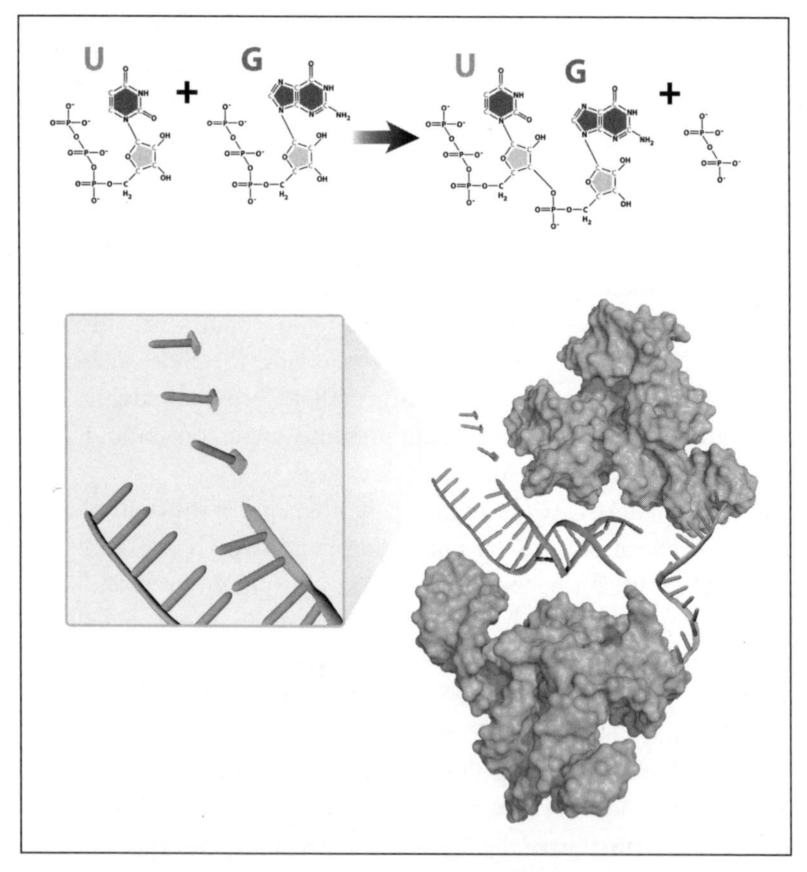

Figure 15.1. RNA replication, by RNA.

of a bound partial helix, thereby extending a growing complementary RNA copy strand at its pointy (3′) end. This is the same reaction as in the top panel, but one now directed by nucleotide base pairing with a template strand.

This nucleotide-joining reaction can be used in SELEX for phosphodiester bond formation, by exploiting the so-called ligase reaction. Ligases join RNAs and DNAs end to end (3′ to 5′), making longer product strands. For example, DNAs are frequently replicated as smaller pieces, then joined into longer complete copies by ligation. SELEX was used to isolate RNA ligases by requiring randomized RNAs to ligate a small sequence to their own ends. No ligation, no replication allowed—because the joined sequence is required for later replication. In this way, one can create a selection for RNA ligase activity. Ordinary RNAs that cannot ligate are left behind, unduplicated. Starting with this selection pressure, SELEX produced several kinds of ligase, one of which made the natural kind of phosphodiester bond when it joined the extra sequence to itself. This one was chosen for further development. This molecule also "ligated" nucleotide triphosphates to itself (as in the upper panel of Figure 15.1, if the left-hand U were instead the end of a long RNA), following base-pairing instructions from an internal template. This selected RNA is therefore carrying out the right reaction, using a built-in template, but it adds only a few nucleotides.

The initial ligase-replicase was made more effective by selecting modified RNAs with new randomized domains on their ends (the smaller domain in the lower panel of the figure), to supply a second jaw for gripping a free primed template molecule (the paired strands in the figure). This audacious idea was successful, in the sense that the evolved multiple-domain molecule can now grip a variety of free primed templates (partial double strands) and can add nucleotides onto the shorter strand fairly accurately, using base pairing to fill in the preexisting notch opposite the template strand (as in the lower part of the figure). In fact, using a particularly favorable primed template, one such evolved RNA could add up to 14 templated nucleotides in 24 hours. This amounts to more than one turn of an RNA helix.

These SELEX reactions show that pure RNAs are capable of a replicase-like reaction, as they would have to be to begin an RNA world with autonomous RNA replicases. However, the available RNAs do not uniformly fill in even relatively short initial RNA templates. This means the existing replicases could not, crucially, replicate their own circa 200-nucleotide length. Nor, for that matter, could they separate the strands of such a long replicated template, to free an initial product strand for further replications. So there are still barriers to conquer in building a ribozyme that would model primordial replicases. Nevertheless, this is significant progress—within the RNA repertoire, free template strands can be copied, using base pairing to nucleoside triphosphates. Thus a crucial prediction of the RNA-world hypothesis is borne out by SELEX experiments.

Those devoted to criticism of the RNA world (or of biological ideas about evolution in general) often point to this as limited progress. One could say of these first tries at an RNA replicase that the replication of 14 nucleotides is not interesting, because no 14-nucleotide replicase activities are known. However, this seems to me an error of emphasis. Looking for a better replicase is a much less uncertain enterprise than wondering if an RNA replicase made of RNA can exist at all. Once we have a replicase activity in hand, it can almost certainly be improved. (See the article by Zaher and Unrau in the list of readings.)

It may be relevant in this connection that, early in these selection experiments, simpler (smaller) selected RNAs were carrying out related replicase-like reactions. And even an unselected nicked short helical RNA with base-paired joinable substrates performs its own nick-joining (ligase) reaction at a low rate—though such spontaneous reactions require an average of about 3 years. Thus, for real replication, catalysts are essential. According to evolutionist Leslie Orgel's first rule, "Whenever a spontaneous process is too slow or too inefficient a *protein* will evolve to speed it up or make it more efficient." For the RNA-world replicase, just substitute "RNA" for "protein."

Readings

"RNA-catalyzed synthesis of complementary strand RNA." Jennifer A. Doudna and Jack W. Szostak. *Nature* 33: 519–522 (1989).
RNA replication by the Tetrahymena ribozyme, in which a complementary strand is assembled from pieces on a template, using the reverse of the self-splicing intron's first step.

"RNA-catalyzed RNA polymerization: Accurate and general RNA-templated primer extension." Wendy K. Johnston, Peter J. Unrau, Michael S. Lawrence, Margaret E. Glasner, and David P. Bartel. *Science* 292: 1319–1325 (2001).
More realistic than the experiment by Doudna and Szostak, because the ribozyme is the product of SELEX (and therefore smaller, though not as small as one would like), uses nucleoside triphosphates (the present universal substrate), and replicates free primed templates with arbitrary sequences, adding up to

14 templated nucleotides. This is a part of the experiment described in this chapter.

"Selection of an improved RNA polymerase ribozyme with superior extension and fidelity." H. S. Zaher and P. J. Unrau. *RNA* 13: 1017–1026 (2007).
This paper requires a stretch to read because of the exotic selection method used, but it is worth the effort because it offers an account of the very best RNA replicase made of RNA that has yet been found, one that can add 20 templated RNA nucleotides.

16

RNA Capabilities and the Origins of Translation

> The meditative eye can look through any single object and see, as through a window, the entire cosmos.
>
> —Aldous Huxley

We should be able to look closely at RNA, in the spirit proposed by Huxley, and see the flicker of an RNA world. To a surprising extent, it has been possible to do exactly this. In this chapter, an aptitude for making proteins—probably veiled for some 4 gigayears—is revealed in modern RNAs.

Bayes' theorem (see Chapter 12) says that a piece of new evidence makes an idea, such as the RNA world, more probable by the exact factor by which the new evidence is more likely to be true if the RNA world existed. For example, the irreplaceable core activities of the RNA organism were surely replication of genetic information and coded translation of genetic information into amino acid sequences (peptides or proteins). Thus, if a ribocyte existed, it had these capabilities. The key point is that we then predict the existence of an RNA-based replicase and an RNA-based translation apparatus. The demonstration that both these previously unknown activities actually do exist (see Chapters 15, 17, and 18) therefore strengthens the case for an RNA world, because it is unlikely that RNA would be found to have all these activities accidentally and independently. To argue to the contrary is like arguing that an arsonist's affinity for storage vessels for combustible liquids is an attempt to anticipate an oil crisis. The experiments on the RNA world discussed in this book are in the same vein—not logical proofs but suggestions, with a high degree of probability, that there was an RNA world, or something much like it.

What then are the reactions whose existence would be sufficient to first make proteins and, eventually, bring on the translational twilight of the RNA world? We can reason by analogy with modern translation, and then circle back to confirm these initial assumptions in light of later experiments. My laboratory and its people have put much effort into this question; see page x for a list of those responsible for each step of the confirmation. Here then is a plausible, if not unique, set of reactions from which to build protein biosynthesis:

1. Amino acids must be made chemically reactive (activated) so that assembly (translation) is the preferred reaction; that is, joining of amino acids is easy because it runs "downhill" energetically.
2. Each amino acid must be attached to the end of the RNA that contains the nucleotide triplet anticode for that amino acid (aminoacyl-RNA synthesis).
3. Aminoacyl-RNAs must react to join amino acids together (the peptidyl transferase reaction).

4. There must be a usable way to predetermine aminoacyl-RNA joining in a specified order (genetic coding).

In order to show that these reactions—none understood as being within the ambit of RNA chemistry at the start of this work—were potentially available to RNA, they have been systematically studied in my laboratory in Boulder, as well as by other biologists. Here are some results, numbered to match the foregoing list:

1. The RNA activation of amino acids, using a reaction that parallels the near-universal biochemical reaction, was found by selection. Because the chemically activated amino acid is highly unstable in normal aqueous conditions (the activated product is stabilized by capture on the surface of a protein in modern cells), a related reaction giving a more stable product and a stabilizing acidity was used for the selection. The result was an RNA that, in slightly acid solutions, made the selected stable product, as well as the unstable activated amino acids leucine and phenylalanine.
2. The synthesis of aminoacyl-RNA was among the first reactions detected by SELEX. The selection supplied the normal form of activated amino acid (the same as in paragraph 1 above) and selected RNAs that transferred the amino acid phenylalanine onto themselves. The selection trick was a chemical reaction that grabs phenylalanine, used to grab the minority active RNA through its phenylalanine in turn. This turns out to be an easy reaction for RNA to perform, and ultimately RNAs were found that were smaller, faster, and more choosy among

amino acids than even modern proteins. Thus, it is highly plausible that RNA performed this reaction at the dawn of translation.

3. The peptidyl transferase has proven elusive to selection. Even today there are no selected RNAs, and no small RNAs, that make peptide bonds via the bona fide peptidyl transferase reaction, that is, that make peptides from aminoacyl-RNAs or their biochemical equivalents. In fact, the search for a small RNA peptidyl transferase has spawned a boneyard of mistakes, false starts, and side reactions. Nonetheless, the argument that peptide bond formation is within the RNA repertoire rests on the most authoritative demonstration imaginable. The natural peptidyl transferase, which is a part of the ribosome, is built into rRNA. Though this modern enzyme relies on proteins to a greater or lesser extent in different kinds of organisms, it is fundamentally an RNA enzyme, and likely a relic of the RNA world itself. The evidence is both convoluted and important, so it has been given its own chapter (see Chapter 17).
4. Coding has been a durable enigma, but it is now clear that a substantial fraction of the code is an expression of the chemistry of RNA interaction with amino acids. Many amino acids are attracted to RNA, which binds them within nucleotide territories that are exceptionally likely to include their coding triplets. This fulfills the hypothesis of a "stereochemical" genetic code, that is, one based on chemical interactions. Presumably there were binding sites for amino acids whose parts were captured for use in the first translation apparatus, thus capturing some of the triplets that we now see in the coding table. Once again, the emergence of the genetic code as an in-

> trinsic expression of RNA chemistry is so important to the RNA world that it has been treated in its own chapter (see Chapter 18). A particularly important result presented in the larger space afforded by that chapter is that some portions of the genetic code may have an entirely different origin.

These four types of reactions, predicted for a translation machine completely composed of RNA, are therefore all detectable, though the full modern parallel for the RNA machine has not yet been assembled. Accordingly, Bayesian reasoning says that it is probable that RNA-world RNAs devised the first translation apparatus (there could be no translation apparatus before RNAs themselves). Because the arrival of translation also initiates the twilight of the RNA world itself, these experiments make it more probable both that the RNA era existed and that RNA's reign as the master biochemical ended as we have reckoned, a result of its own cleverness in translation.

Finally, the success of RNA in taking on new, previously unknown (translational) functions should be viewed in the same light as the indications that life adopted a chemistry prevalent in solar nebula (see Chapter 6). The successes of SELEX show that, hidden in modern pure RNAs, there is a broad chemical bridge leading to RNA-performed translational reactions, though all such connections have lapsed in a modern nucleic acid–protein organism. No sensible person would have confidently bet on these RNA capabilities before selection experiments revealed their existence. Therefore even as there is a hidden chemical bridge to synthesis of varied amino acids, bases, and nucleosides, a similar bridge in the form of unexpected translational capabilities has been found, latent and presently unused, in the RNA molecules of today. This is exceedingly unlikely to be an accident, especially when all the newly selected, essential reactions are considered together.

The ready appearance of this missing connection to the history of life on Earth is therefore a gratifying surprise. As one crucial result, we can move, at least in thought, one full step back toward the beginnings of earthly life. In dimly seeing our ribocyte ancestor's invention of coded protein synthesis, we put in place substantial evidence of an RNA-dependent step during the early history of life on Earth. As a result, grasping the origin of translation has an unavoidably eerie feel—we are perceiving and partially recreating otherwise imperceptible events, lost in deep time.

Reading

"On translation by RNAs alone." Michael Yarus. *Cold Spring Harbor Symposia on Quantitative Biology* 66: 207–215 (2001). *Reviews the biochemical reactions required for translation in light of known and selected RNA capacities. An expanded form of the argument presented in this chapter.*

17

The Quest for the Peptidyl Transferase

To create something you must be something.

—Johann Wolfgang von Goethe

As might be hoped, from the constitution of the peptidyl transferase, the enzyme that makes all the peptide (amino acid–amino acid) bonds in every earthly protein, we can read some profound lessons for the invention of protein synthesis in the RNA world.

We start with model chemistry, performed by the ribosomal particle (the subcellular arena for translation) with simplified reactants. By using fragments of normal molecules, a simpler biochemical model (called the fragment reaction) can be used to emulate the central reaction of the ribosome, that mediated by the peptidyl transferase. As an example, the fragment reaction can be used to show that the ribosomal particle, which comes apart into two major pieces, each about half the size of a very small virus, has the peptidyl transferase built into its larger piece, the so-called 50S particle.

Using a similar model reaction, Harry Noller of the University of California Santa Cruz had shown that peptidyl transferase activity had unexpectedly little dependence on the proteins of the ribosome, which stud the surface of the rRNA like the chips on an ovoid chocolate chip cookie. In the Noller laboratory, it had been shown that rRNA exposed serially to three different treatments designed to strip off the proteins still retained the ability to make peptide bonds via the fragment reaction. Therefore it was easy to imagine that the chips (proteins) were minor participants.

This observation might seem to close the case, but surprisingly, the triply denuded rRNA molecules still retain 3–6 chocolate chips, ribosomal protein molecules, which they apparently protect from removal. When these proteins *are* finally removed, the fragment reaction stops working. Thus either the RNA is doing peptidyl transfer, and requires only a few proteins, or a small minority of proteins might be the peptidyl transferase, or both molecules might be getting into the peptidyl transferase act.

These possibilities were resolved only by determining the atomic structure of the ribosome. This is an amazing story, beginning with the means for the structural determination. To resolve structures at the atomic (crystallographic) level, one must line up essentially identical molecules in three dimensions to make highly ordered crystals. For well-aligned molecules, the effect of the crystal on a beam of X-rays manifests the details of the underlying identical molecules. This is an easy enough procedure with a small molecule, but it is a remarkable feat for the ribosome, which contains more than a hundred thousand atoms. Crystallization of the ribosome was shown to be possible by Ada Yonath at the Weizmann Institute of Science in Rehovot, Israel. The first high-resolution ribosome structures came from groups led by Peter Moore and Tom Steitz at Yale University. Both the Yonath and Steitz groups were recognized (along with Venki Ramakrishnan of

Cambridge, England) with the 2009 Nobel Prize in Chemistry for their work on the structure of the ribosome.

Within these structures the active center at which amino acids are joined (the peptidyl transferase center) was located using synthetic molecules designed to resemble the starting materials, as well as a molecule resembling two amino acids caught in the process of forming a peptide bond. Because the peptidyl transferase necessarily has a pocket for forming bonds between amino acids, the latter synthetic molecule (called CCdApPuro) will sit stably in the ribosomal pocket. Thus when CCdApPuro (synthesized by Mark Welch in my laboratory) was allowed to diffuse into crystals of the large ribosomal subunit at Yale, the site where it came to rest accurately marked the site of peptide bond formation. The relevance of this story is that this site was just where it should have been. For example, it was at the head of a tunnel so the newly linked protein could exit the ribosome. Even more notably, it was in a neighborhood composed of RNA, with proteins relatively far away. The modern peptidyl transferase had to be RNA, and potentially closely related to the original peptidyl transferase of the RNA world.

Despite intensive study and the accumulation of much more critical information, this is the picture we retain today. Every regular bond between amino acids in proteins on Earth is forged in a crucible made entirely of RNA. Ribosomal proteins help to hold the peptidyl transferase region together in today's ribosome, but the joining reaction is conducted by RNA, accelerated by RNA groups that reside on reacting aminoacyl-RNAs themselves, as well as, to a smaller extent, some parts of the rRNA. Thus the idea that peptides were first encoded in an RNA world is supported by the finding that peptidyl transferase within the ribosome is still composed of RNAs, including bits of both rRNAs and tRNAs. However, there is an even more direct argument.

SELEX experiments with CCdApPuro were performed 5 years before the X-ray crystallographic work with the large ribosomal subunit. The synthetic compound can be linked to a small agar bead, represented here by ⊗, to make ⊗-CC-dApPuro. These beads allow selection of RNAs that bind CCdApPuro—one can pick RNAs that were attracted to the decorated bead by grabbing anything that comes along with the beads (see Chapter 14). The two most abundant RNAs that bound CCdApPuro had a similar sequence of 8 RNA nucleotides adjacent to the CCdApPuro: AUAACAGG. It was electrifying to find, 5 years later, the same sequence of 8 nucleotides next to bound CCdApPuro in the real ribosome. There are 65,536 different sequences of 8 nucleotides, so it is unlikely that this sequence would be the same in all three places by accident. The apparent implication is startlingly direct: the RNA cradle that holds reacting amino acids in peptidyl transferase will arise in one step from a random RNA sequence if a structure that resembles reacting amino acids must be bound. This apparent duplication of the evolutionary origin of the core RNA sequence of peptidyl transferase substantially strengthens the idea that translation, as performed in every modern organism, originated in the RNA world.

But there is even more information to be extracted by considering the crystallographic and selection results together. The role of the AUAACAGG sequence that appeared in the binding loop of both of the selected RNAs (as well as in real ribosomes) is known from later ribosome crystallography. This means that we can deduce other things about the origin of translation. For example, the role of this sequence on the

The predictability of evolution

There is another way to look at the isolation of the core octamer of the peptidyl transferase from random RNA sequences, using affinity for CCdApPuro. Because the two most probable solutions to this selection both employed the core Welch octamer to bind CCdApPuro, this suggests that, at least at the time of the origin of peptidyl transferase, evolution proceeded by deterministic routes that can be reproduced on today's laboratory benchtops. This contrasts with the view that evolution contains a large component of chance that would rule out the same outcome if "the tape were played again." Because the chance notion has gotten so much press, it is worth mentioning that evolution, at least in simpler times, can have a predictable molecular outcome.

ribosome is to help align aminoacyl-RNAs for peptide bond formation. That means that aminoacyl-RNAs were the reactants used to make peptides at the origin of the peptidyl transferase, as they are now. It means that RNAs must have looked then much as they do now (had much the same chemical structure), or else the sequence selected today would not be the same as the sequence selected then. That CCdApPuro is an effective inhibitor of ribosomes from all three domains of life means that the peptidyl transferase, as we would expect, predates the separation into life's current three domains. That is, peptidyl transferase dates from before the LUCA—just as the above argument tracing it to the RNA world predicts.

Remarkably, the results of the ⊗-CCdApPuro selection also reflect back on the assumptions made at the start of our discussion of translation. Because the AUAACAGG sequence interacts with aminoacyl-RNA (or at least aminoacyl-nucleotides) and dates from the appearance of peptidyl transferase in its modern form, the aminoacyl-nucleotides have probbaly been used for translation over the same period. This makes it probable that the syntheses of aminoacyl-RNA and peptidyl transferase are old reactions, on the central route to peptides from the beginning of protein biosynthesis, as we assumed at the start of Chapter 16.

Readings

"The structural basis of ribosome activity in peptide bond synthesis." Poul Nissen, Jeffrey Hansen, Nenad Ban, Peter B. Moore, and Thomas A. Steitz. *Science* 289: 920–930 (2000). *The first view of the peptidyl transferase (and the large ribosomal subunit) in molecular detail.*

"Peptidyl transferase: Ancient and exiguous." Michael Yarus and Mark Welch. *Chemistry and Biology* 7: R187–R190 (2000). *A short review on the consequences of the identification of the biological peptidyl transferase on the 50S ribosomal particle.*

18

A Language Much Older Than Hieroglyphics: The Genetic Code

A different language is a different vision of life.

—Federico Fellini

The title of this chapter, a phrase from the book *The Language of Life* by George and Muriel Beadle, evokes the role of the genetic code, which records the relations between the mRNA language (triplet nucleotide anticodons or their equivalent, complementary, antiparallel codons) and the protein language (amino acid sequences). In so doing, the code determines the meaning of the 1.2% of human DNA that implies the sequence of structural and enzymatic proteins. These definitions not only predate ancient Egypt, they come from an RNA world that probably flourished and died while the planet Earth was truly young.

As an example of the coding of these interlanguage relations, UGU triplet codons (3 adjacent nucleotides) in mRNA mean the amino acid cysteine in protein emerging from a ribosome. All ribosomes are probably incorporated in this case, because I have chosen an example not known to differ in any variant genetic code. There are 63 more of these triplet nucleotide–to–amino acid relations that make up the complete code, and they are listed in Figure 18.1. Treating the code in this way, as a translation between languages, suggests the notion that variation in the two languages was possible. Though not always mentioned, it is a fact that our cells could be speaking another coding dialect than we do, like the difference between the utterances of francophones in Montreal and Paris. Only sometimes might RNA triplet UGU mean amino acid cysteine. But though slightly varying dialects exist in organisms, and particularly inside their organelles (subcellular compartments), the code is not unreservedly free to vary.

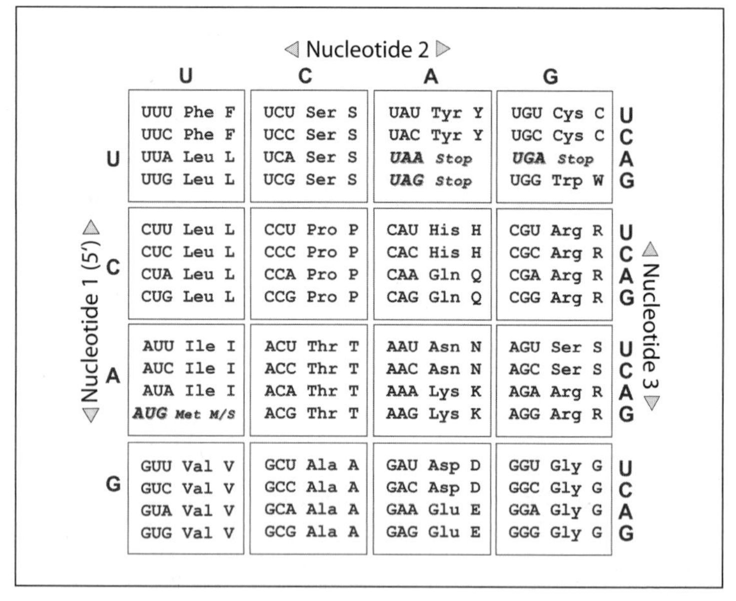

Nucleotide 1 (5′)	Nucleotide 2: U	C	A	G	Nucleotide 3
U	UUU Phe F	UCU Ser S	UAU Tyr Y	UGU Cys C	U
	UUC Phe F	UCC Ser S	UAC Tyr Y	UGC Cys C	C
	UUA Leu L	UCA Ser S	***UAA*** *Stop*	***UGA*** *Stop*	A
	UUG Leu L	UCG Ser S	***UAG*** *Stop*	UGG Trp W	G
C	CUU Leu L	CCU Pro P	CAU His H	CGU Arg R	U
	CUC Leu L	CCC Pro P	CAC His H	CGC Arg R	C
	CUA Leu L	CCA Pro P	CAA Gln Q	CGA Arg R	A
	CUG Leu L	CCG Pro P	CAG Gln Q	CGG Arg R	G
A	AUU Ile I	ACU Thr T	AAU Asn N	AGU Ser S	U
	AUC Ile I	ACC Thr T	AAC Asn N	AGC Ser S	C
	AUA Ile I	ACA Thr T	AAA Lys K	AGA Arg R	A
	AUG *Met M/S*	ACG Thr T	AAG Lys K	AGG Arg R	G
G	GUU Val V	GCU Ala A	GAU Asp D	GGU Gly G	U
	GUC Val V	GCC Ala A	GAC Asp D	GGC Gly G	C
	GUA Val V	GCA Ala A	GAA Glu E	GGA Gly G	A
	GUG Val V	GCG Ala A	GAG Glu E	GGG Gly G	G

Figure 18.1. The genetic code. In this common representation, the three nucleotides of the coding triplets are associated with three- and one-letter abbreviations for the 20 frequent biological amino acids.

The Genetic Code

Codons are shown in the boxes in Figure 18.1, alongside three-letter abbreviations for the amino acids (e.g., Phe = phenylalanine) and one-letter abbreviations for them (e.g., Phe = F). Stop signals are depicted in italics, as is the start signal (which also means Met [= methionine] in the middle of a protein) shown at the lower left. The left-edge UCAG indicates the first nucleotide of the codon (5′), the top edge shows the second or middle nucleotide, and the right edge shows the last (3′; possibly ambiguous or "wobble") nucleotide. Thus CAU/CAC are codons for histidine (= His = H). *Stop* is the termination signal. The figure may also be useful as a mnemonic and a list of the standard 20 amino acids and their common abbreviations.

The slight variation actually observed in the code contrasts with the variation that might exist. With $4 \times 4 \times 4 = 64$ codons or anticodons and 20 usual amino acids (plus stop) to encode, if we pick amino acids and randomly associate them with codons (giving each amino acid plus stop at least one codon), there will be 1.5×10^{84} ways to associate amino acids with the codons that translate them. Even if many codes cannot appear because some amino acids and their coding triplets are interconnected by chemical principles (as discussed later in this chapter), there are still a vast number of possible codes. Modern creatures use only a highly related few from among this vast range. As we have said, this convincingly implies that genetic coding in different organisms cannot possibly come from independent selection of different codes, but instead has the unity expected of an inheritance from one ancestor, the LUCA. Thus, alongside the similar basic energy-generating biochemistry of organisms and the similar nature of biosynthesis within organisms, the code is one of the pillars of the argument for the remarkable, indeed dazzling, original unity of modern life on Earth.

Genesis of the Code

So where did the genetic code come from, and when did it appear? There can be no code until there are reproducible RNA sequences, because one side of the coding equation(s) is particular RNA triplets, the codons. The word "reproducible" is worth some emphasis; it suggests the notion of accurate replication. There can be no effective coding until RNA sequences can be duplicated and preserved. Thus, in addition to our usual deduction of the necessity for early replication in order for Darwinian evolution to begin, genetic coding is not plausible until the same sort of accurate replication is established. This jibes with other basic assumptions: replication initiates the RNA world, and the code arises afterward with accurate replication as a foundation, appearing as a favorable elaboration of the doings of some ribocytes. The genetic code is an invention of the middle RNA world.

Polymers made of amino acids (peptides, proteins) long ago won the evolutionary race to carry out the majority of biological chemistry. Therefore polymerized amino acids are, on the whole, superior to polymerized RNA nucleotides as cellular chemical agents. We can, without extrapolation, regard wielding chemically active amino acid sequences as an advantage for a ribocyte. Thus, the driving force for the invention of translation (RNA-instructed polypeptide or protein synthesis) is not a mystery. Rather the question is how translation and the code might have first appeared in an RNA world that could reliably duplicate an RNA sequence.

Francis Crick once considered the notion that codons and amino acids became arbitrarily associated (calling the code a "frozen accident"). However, an alternative idea posits a physically realistic interaction that might precede the code and give rise to a first draft of it. These interactions would define a primordial set of RNA–amino acid relations that can grow into the complete code we find throughout Earth's biota. This is called the stereochemical hypothesis, a name that evokes the complementarity in shape and chemistry between the primordial RNA sequence and amino acids that are envisioned associating with it. These associations would not necessarily all be preserved in the modern coding table (perhaps being overwritten by later changes), nor would they necessarily embrace all 20 modern amino acids (perhaps just a founding subset that has survived). However, they would supply a starting point.

Searching for the historic stereochemical interaction at first seems relatively easy. If chemistry it was at the base of the code, then we just need to repeat the right experiment to see it recur. Chemistry at the time of origin should be the same as chemistry now; the chemical interaction should be easily reproduced on a modern laboratory bench. Yet there are many possibilities: probably the code began with a subset of favorable examples—a plausible subset being the set of amino acids that interact readily with RNA. But which ones, and how did that set expand to the complete code we know today, in which all 64 coding triplets have an explicit meaning, as amino acid or stop or start (Figure 18.1)? And what was the founding interaction, which we might hope to reproduce?

Early attempts to demonstrate such interactions found nothing interesting. For example, Paul Doty at Harvard University confined rRNA within a bag permeable to amino acids and asked if amino acids would enter the membranous bag to bind to the RNA. The RNA did not seem to attract amino acids to bind to it. As we shall see, though Doty's idea was correct in some ways, amino acid binding sites are too rare to be detected by this experiment.

The first usable clue came, as it so often does, from an experimental avenue that seemed unrelated. The group I self-splicing intron, which removes itself from a *Tetrahymena* rRNA precursor and seals the ends of the resulting gap together, was under study in Tom Cech's laboratory (see Chapter 11). The enzymatic cutting and sealing activity was clearly entirely within the intron–intervening RNA sequence, because the intron RNA retained the ability to perform chemical reactions after it had cut itself out of the initial rRNA precursor. The intron started by reacting with a free G nucleoside or nucleotide to make the first nick. The intron contained hundreds of nucleotides of intricately folded RNA, and accelerating chemical reactions requires bringing reacting groups together with atomic precision. The group I intron therefore potentially reported its detailed shape via its ability to perform the splicing reaction. It was a complex RNA structure that could detectably respond when an amino acid had bound to it (and slightly changed its shape) through a change in the rate of its own splicing reactions. Therefore the binding experiment was to perform self-splicing reactions in the 20 standard amino acids and see if any of them affected the reaction.

In fact, in my laboratory we were very excited to find that only one of the "magic 20" biological amino acids slows *Tetrahymena* self-splicing RNA reactions. Further study of this inhibition (by the amino acid named arginine = R) showed that instead of binding generally to the intron RNA, arginine exerted its effect by entering and blocking the RNA site for the G (guanine) nucleoside splicing initiator. No G can be bound to the arginine-RNA because of blockage by the previously bound arginine; hence no self-splicing. The consequent splicing slowdown signaled amino acid binding, as we had hoped. The binding and inhibition by arginine had all the earmarks of a normal molecular interaction; for example, the *Tetrahymena* intron's G-arginine site detected the difference between right-handed and left-handed forms of the amino acid.

It therefore appeared that RNA could bind and discriminate amino acids. RNA–amino acid interactions like those envisioned as the foundation of a stereochemical RNA–amino acid code could now be studied. In fact, the *Tetrahymena* RNA affinity for arginine could be understood as a molecular resemblance between a piece of arginine and similarly placed atoms in the G splicing cofactor, for which the self-splicing RNA necessarily has a binding site. But this site is a peculiarity of the self-splicing RNA; more general binding results were needed—and were just ahead.

The G-arginine binding site of the group I intron was soon located in the *Tetrahymena* RNA sequence in experiments by Francois Michel in Jack Szostak's laboratory. In fact, just where we now expected G-arginine to be sitting, we could now see two codons for arginine, one of them highly conserved throughout all known group I RNA sequences! The suggestion was clear, electrifying even now, years later—an amino acid had an affinity for (some of) its own codons, not as free nucleotides or even free codon trinucleotides, but as a part of a larger RNA structure, within which the coding triplet is an essential part of a site with a shape favorable for binding the amino acid.

These ideas were reinforced by SELEX experiments (see Chapter 14). Around 1990 three laboratories under the leadership of Larry Gold, Gerry Joyce, and Jack Szostak virtually simultaneously devised methods for the isolation of new active RNAs from large groups of mostly inactive molecules. Because "large" here means 10^{15} molecules or so, the technique emulates very effective evolution to isolate RNA with activities that may have existed in ribocytes, but that have no modern biological equivalents. SELEX has been important in exploring characteristic ideas of the RNA world, as we have seen at several places in this book.

The basic idea is uncomplicated in principle; among randomized RNA sequences with no preordained sequence or structure—made by transcription (RNA synthesis) from DNA with every template nucleotide a mix of all four—there will be a tiny minority of molecules with a new activity. Given a way of physically separating the minority active molecules, these can be purified, cloned, and studied. Thus desired or predicted or essential new RNAs can be studied without having to know what they might look like, and without having to find them in a wombat or a wildebeest. This ability is crucial for the many molecules that once might have served ribocytes, but that probably vanished from the Earth gigayears ago. (The method is described in more detail in Chapter 14 and also in Chapter 7.)

Now we can return to the experiments on the genetic code. With the *Tetrahymena* self-splicing RNA results in mind, we first wanted to see if arginine binding sites were abundant among RNA sequences. In fact, binding proved easy to isolate; the natural arginine site in the *Tetrahymena* intron was not a fluke, but a representative of an easily reproduced capacity of RNA, readily extended by finding new examples. A particularly interesting arginine site, isolated by Michael Famulok of the University of Bonn, was selected by beginning with a previously selected site binding citrulline, a similar chemical. When this

RNA was required to convert from binding citrulline to binding arginine, it acquired three nucleotide mutations. The three changes created two new arginine codons. When these were located within the three-dimensional structure of the RNA binding site, the changed nucleotides (and therefore the two emergent arginine codons) were in close contact with the bound amino acid arginine (Figure 18.2). In this case, one could actually see how folded RNA sequences containing coding triplets

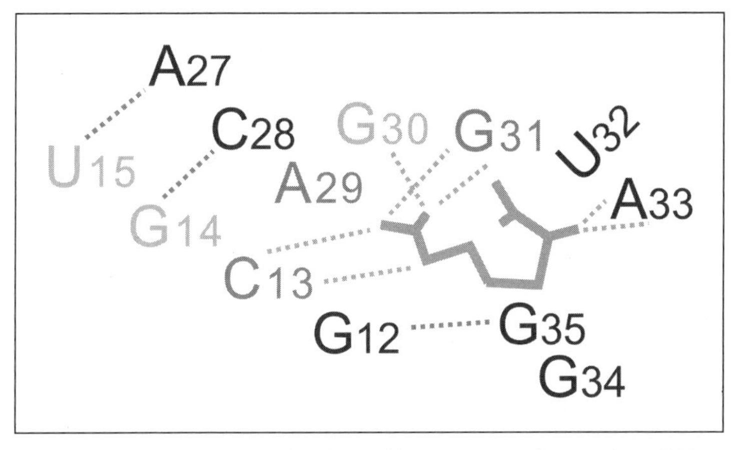

Figure 18.2. An amino acid embraced by cognate coding triplets. RNA backbones are not drawn but can be inferred by following the numbers on successive nucleotides. The brief U=A G≡C helix at top left expands into a 29-35 loop that cradles the bound arginine, which is the light gray stick structure in the middle. Dashed lines are internucleotide or RNA-amino acid interactions. The essential point: nucleotides C13, A29, and G31 changed when arginine binding was selected (from citrulline binding). All three changes participate in both the arginine binding pocket and new A29-G30-G31 and C13-G14-U15 arginine codons. Based on Y. Yang et al. "Structural Basis of Ligand Discrimination by Two Related RNA Aptamers Resolved by NMR Spectroscopy." *Science* 272: 1343–1347 (1996).

could make up part of a binding site for the amino acid, when present as a part of a larger overall structure.

This kind of selection experiment has since been greatly extended. Even without a three-dimensional structure, the RNA nucleotides involved in binding an amino acid can be determined by chemical methods. For example, the nucleotides protected from a small chemical reagent when the amino acid is supplied are likely to be in the RNA binding site or linked closely to it. If one groups all such nucleotide sequences, they can be compared by chemical criteria with sequences in the same RNA but unlinked to the binding site. The latter nucleotides have arisen by the identical experimental route as the binding nucleotides, but they probbaly have less important functions in amino acid affinity. Then one can ask: do the nucleotides in the site (and therefore close to the bound amino acid) have more coding triplets than the nucleotide sequences in the same molecule but without a close connection to binding?

At the time of this writing, 8 amino acids of 20 have been examined by SELEX. We have data for 349 amino acid binding sites whose molecules contain 18,513 RNA nucleotides that can be examined for coding triplets. These RNAs bind water-hating amino acids like phenylalanine and isoleucine, as well as those with water-loving structures, such as histidine, arginine, and glutamine. The net result is that both codon and anticodon triplets are unexpectedly concentrated in the sites for their cognate amino acids. Of the 8 amino acids tested, 6 show some RNA triplet–amino acid association. As originally suggested by the natural example of arginine and *Tetrahymena* RNA, codons for arginine are concentrated in the selected arginine sites. But anticodons for arginine also appear when this many sites are inspected. Only anticodons associated with phenylalanine appear abnormally frequently in phenylalanine sites. Both codons and anticodons appear in isoleucine sites. There is a complete negative, itself interesting: amino acid glutamine sites display neither nucleotide codons nor anticodons at unexpected frequency. There is other evidence that glutamine entered the genetic code by another route, called coevolution, that has long been favored by J. T.-F. Wong of Hong Kong University. In this mechanism, metabolically related amino acids inherit some of the codons for their most ancient related biochemical precursor, when it becomes possible to produce the later compounds. If glutamine is one of these amino acids, asparagine (a related member of the 20) will probably also be one.

To return to the stereochemical code: overall (counting positives and negatives together), there is only one chance in 10 trillion that codons are arbitrarily scattered about with respect to their binding sites, and one in 10 trillion trillion that anticodons are similarly unassociated with binding site nucleotides. So the association first seen in the self-splicing RNA has been borne out in a large sample of newly selected binding sites. In fact, the newly selected arginine sites can be used as a rather good argument on their own: there is only one chance in 30 million that newly selected arginine binding sites do not concentrate arginine codons. Overall, considering all examined triplets together, positive and negative cases accounted for, there is only one chance in 2×10^{36} (two trillion trillion trillion) that coding triplets and amino acid binding sites are unrelated in the complete current population of selected sites. This is exceedingly strong evidence from SELEX, far beyond normal scientific standards for evidence against a null hypothesis (a hypothesis of no relation). The initial hint from *Tetrahymena* RNA has proven broadly, surprisingly valid—the genetic code preserves a fundamental relation between coding triplets and RNA–amino acid interactions.

There is another kind of indication that the hypothesis may be correct. One can focus a selection experiment on simpler sites, amino acid sites containing fewer nucleotides. These simplest, smallest sites would also have been the easiest sites to find in primordial conditions that did not easily allow large RNAs. When this is done for isoleucine, histidine, and tryptophan, as well as arginine, the smallest, most abundant sites in all four cases have the property of displaying at least one conserved (required) anticodon or codon. Thus if, at the time of the code's formation, the most readily found RNA structures were used to bind amino acids, the structures most likely to have been elected would have automatically contained coding triplets and their particular amino acids in close physical apposition. This tendency is robust, observed far more often than any plausible accident. So it is likely that the code originated with RNA-bound amino acids, when pieces of their binding sites were captured to serve as codons and anticodons as a modern translation apparatus evolved. The structures found originally at the heart of the self-splicing RNA may be "molecular fossils," fragments of the authentic origin story for the Earth's genetic code. This overall scheme has been called the "escaped triplet hypothesis" to commemorate the notion that coding triplets first arose in amino acid binding sites and later escaped to ply their modern translational functions.

The Rest of the Code

There are certainly plausible alternative histories for the genetic code. The coevolution theory, already mentioned, emphasizes a simpler original code with fewer amino acids. As biochemistry becomes more competent, newly synthesized amino acids related to those of the original code take over some of the original codons used by their amino acids. This is probably true for amino acids like glutamine (= Q) that bind poorly to RNA when free and show other RNA evidence of coevolution.

A third alternative, the adaptive theory, suggests that the code has evolved to minimize the impact of likely mistakes by giving related codons to amino acids whose structure (and whose use within a protein superstructure) is similar. This is a viewpoint most convincingly worked out by Steve Freeland of the University of Maryland, who has shown that the genetic code minimizes the chemical consequences of amino acid mistakes. The adaptive theory may be true, for example, of a subset of the hydrophobic (water-hating) amino acids that are poorly distinguished by RNA but that have essential roles in proteins, forming the core of any larger protein structure.

Both adaptation and coevolution suppose an ancient core of assignments made on some other basis, and, in common with stereochemistry, both imply that related amino acids will have related codons. Therefore, adaptation and coevolution easily coexist with a fundamentally stereochemical origin for the code. There may well have been serial stereochemical and adaptation or coevolution eras for a particular amino acid. In fact, there appear to be modern biochemical remnants of coevolutionary expansion of the code: amino acids that are synthesized on tRNAs so that they acquire the corresponding codons. Similarly, the real code is highly organized according to apparent amino acid similarities, so adaptation also appears to be a plausible guiding factor in some of the code.

The most likely view of the history of the code seems to be that it originated stereochemically, then codons came to their present assignments by all three routes. Because 6 of 8 amino acids show significant traces of stereochemical assignments, the stereochemical era may have seen the entry of most modern amino acids into the code. Because only about 1 of 4 potential triplets shows concentration in binding sites, it may nevertheless have been that most triplets entered the code by other means. In other words, RNA–amino acid interactions may have supplied the working material for expansion of the genetic code via adaptation and coevolution. Such a combined history seems at this point to explain all observations.

It sometimes strikes commentators as surprising that all earthlings are the descendants and inheritors of one code, and therefore the descendents of a LUCA. It may be that the breakthrough organism that invented the genetic code had such an evolutionary edge that others never caught up. However, in the sufficiently long run it is automatically likely that only one group of organisms survives. In the sufficiently long run, there is automatically one code and one line of descent alive to muse about its history. Extinct creatures took their codes into the abyss, along with any possibility of examining the intermediate coding experiments that might have contended with ours. So the only clearly meaningful questions we can ask are solely about our code and our own ancestors. However, even though the genes of many thousands of different types of organisms have not revealed any independent alternative codes, there are still more codes and organisms to see, even on Earth—and perhaps elsewhere in the solar system.

Finally, we can say that the RNA at the time of the invention of the code must have been much like ours, because small changes in the structure of nucleotides disrupt the complex structures that alone have complex RNA functions (see Chapter 10). Our present RNA still records the formation of stereochemical coding assignments because modern experiments reproduce some of the code. Therefore the RNA that existed at the appearance of translation (the middle RNA world, when RNA sequences were first available) must, amazingly, resemble the RNAs of our time, some 4 billion years later.

Readings

"The genetic code is one in a million." Stephen Freeland and Laurence Hurst. *Journal of Molecular Evolution* 47: 238–248 (1998).
The paper that launched the modern consideration of the possibility that the code was shaped by the need to minimize the consequences of errors (the adaptive theory).

"Coevolution theory of the genetic code at age thirty." J. T.-F. Wong. *Bioessays* 27: 416–425 (2005).
A discussion of the coevolution theory of the shaping of the code, by the major exponent of this route to the code's architecture.

"Genetic code origins." Michael Yarus and Eric Christian. *Nature* 342: 349–350 (1989).
The first intimation of the association between amino acid binding sites and coding triplets.

"RNA–amino acid binding: A stereochemical era for the genetic code." Michael Yarus, Jeremy Joseph Widmann, and Rob Knight. *Journal of Molecular Evolution* 69: 406–429 (2009).
A current summary of the evidence for RNA–amino acid interaction as the source of the initial code (the stereochemical theory).

19

Assume a Spherical Cow: The Ribocyte

Every man is a quotation from all his ancestors.

—Ralph Waldo Emerson

Finally we meet the beings we have been seeking. What can we say about our predecessors from the RNA world? Let's call them "ribocytes" or RNA cells. (I'm also tempted to call them "ribosaurs," a joke suggested by Peter Moore, whose participation in the determination of ribosome structure was mentioned in Chapter 17.) As you can anticipate, speculation and probability necessarily become very prominent in any such discussion.

Ribocytes were dependent on RNA by definition. It is also likely they were cells, because being membrane-delimited is by far the most plausible way to maintain a coherent evolutionary identity, so that there is a "you" who can benefit from your own chemical talents. The outer membrane that defines the limit of a cell can be made of cosmochemicals, that is, amphipathic lipids (lipids having both a water-loving and a water-hating end) which (again!) occur naturally and broadly in the universe and do not require complex biosynthesis. Therefore, in this context, we take a cell membrane made of such lipids as an easy, early, natural structure, and consider our problem to be the accumulation of the other necessities for life. You may want to think about the writings of people who argue for an alternative surface origin for life, in which a patch of mineral surface was a Darwinian individual, instead of the volume on the interior of a membrane. However, this has always seemed to me much more difficult to rationalize chemically, so our ribocytes or ribosaurs remain membrane-bounded volumes.

So what would we have to put inside our cell membrane to have a creature that might populate an RNA world? As we have seen, we require machines for replication and for translation, and because both these activities require energy we also need an energy-generating metabolism to fuel our ribocyte. We may also need other tricks, such as assembly lines for nucleotides to feed replication, but these may have been supplied externally, and if so would be less urgent.

Where can we get information about these topics? We have the only outcome of our selection experiments, as well as the chemical properties of RNA itself. We are in the position of guessing about events 4 billion years gone, using scattered observations. As you will appreciate, such a long arm of speculation reaching out from such a narrow base of facts makes it likely that we cannot point very precisely. Nevertheless, we can offer some informed speculations.

The first thing to grasp—and the rich implications of this finding seem to be seldom appreciated even by professional biologists—is the fact that RNA has turned out to be so versatile. There was a time when it was thought to be an inert trans-script, a stand-in for the "real information" in DNA. But RNA in the modern view carries information, regulates genes, makes peptide bonds, and plays a central role in modification of coded RNA information itself (see Chapter 9). It is truly indispensable. Yet a ribocyte would have required much more, because it would have been required to do everything with RNA. In an age when the only large molecule was RNA (or something very much like it; there was as yet no DNA or protein) there would have been no alternative.

The fact that an RNA world required RNA to have many functions, now lost, and that these functions have been substantially recreated using SELEX (see Chapter 14), is a powerful, many-armed argument for the RNA world hypothesis.

Ribocytes were s-l-o-w. One might expect newly selected catalysts to be less efficient than highly evolved current cellular ones, which have been perfected over gigayears. This idea is borne out by newly selected RNA catalysts from SELEX experiments. Because these selections, in common with evolution itself, only find the reactions most easily found, and not the fastest ones, the earliest ribocytes would have been constrained to do things s-l-o-w-l-y. Measurements on recently evolved RNAs suggest that RNAs that are easily found by selection might react (approximately and variably) 10 million times slower than rapid, highly evolved protein enzymes. We can appreciate the consequences of this observation by considering cell multiplication. Cell division requires catalysis of cellular duplication, and no cell can divide faster than it can reproduce all its essential molecules.

This is clearest with regard to genetic information. Recent experiments suggest that the minimal information for capable ribozymes would consist of 40 or so nucleotides, about 15–20 of which might be essential and highly conserved. These dependable elements combine to make the RNA active site, where chemistry occurs. We might guess from the study of selected RNA catalysts for replication that reproduction of such units would have required some days. Thus it is not possible to conceive of a ribocyte, relying on RNA catalysis to reproduce the simplest active RNA structures, that would be ready to reproduce in less than days. This is much more leisurely than the minimal 10–20 minutes required by the fastest modern cells. Therefore, ribocytes would have grown slowly by current cellular standards.

However, ribocytes cannot grow too slowly. RNA itself is chemically unstable; it cannot take longer to reproduce than it does to decay, or it will not persist. Because we need relatively stable inheritance for Darwinian outcomes, we must ask how long the smallest RNA-like units (genes or catalysts) would last. Supposing that 20 inter-ribonucleotide linkages must survive, the answer (in neutral solutions at room temperature) is some tens of weeks. Therefore, a dividing ribocyte is caught between the limits of its own slow catalysis and the instability of its superstructure—it must divide every few days to every few weeks. While these conclusions rest on many assumptions, they are based on experiment and so can be modified if we consider a world that is, say, hot rather than like an indoor laboratory with central heating and cooling.

Slow they may be, but ribocytes must also be relatively complex (by comparison with other natural objects). This conclusion is forced on us by consideration of the simplest activities that we can imagine them carrying out. To see that ribocytes could be smaller than modern cells, we need to do some numerical fiddling.

As before, a defensible RNA world begins with the invention of RNA replication. We are forced to suppose that translation and a supporting metabolism existed, at least near the end of the RNA world. If we put together the initiation of the RNA world (by replicators) with the dusk of the RNA world (in the time of the early translators), we find that we are considering a ribocyte that could make and maintain unstable large molecules like the RNAs, could create the activated monomers needed for replication (nucleotides), and could concentrate amino acids and attach them to RNAs so that translation (coded by RNA sequences) could be managed. All these processes require energy and therefore a competent energy metabolism.

Therefore, we cannot imagine the RNA world without supposing that ribocytes could do much more than we have said so far. In fact, there are many ancient biochemical reactions they might have possessed, as treated in the supplementary readings listed at the end of this chapter. The simplest modern microbes, observed in detail through their sequenced genomes, have a few hundred essential genes. Therefore a notably more primitive ribocyte with perhaps a hundred genes seems a plausible hypothesis, and it might still carry out most essential reactions, given a starting soup. One hundred genes at 40 nucleotides per function is 4 kilobases of unique sequence in the genome. In order to meet the schedule for division mandated by RNA stability, each of the 100 probably has to have a more or less committed replicase, which must replicate its specific gene sequence, then replicate the replicase itself. This back-and-forth is required so that a new cell can have all its specific functions, as well as the replicases required to maintain them, about a week after the previous division.

Mechanisms that partition cellular contents into two daughter cells created by division, to prevent creation of defective progeny, are presently very intricate. Ribocytes probably did not have such carefully calibrated partitioning. Instead any given bit of the parental cell's interior is probably distributed randomly—the probability that a given RNA goes to either daughter is 1/2. If there is one copy of each of our 100 hypothetical essential RNAs, only a minute fraction,

7.9×10^{-31}, of the randomly provided daughters has everything it needs after division. If there are 5 copies of the 100 RNAs, 4% of daughters get all functions. This is still not enough for cell division to regularly yield more than one viable cell. But the solution to the problem comes easily: if 10 RNA copies exist, then 91% of daughter cells are completely equipped with the 100 essential RNA functions, even after random divisions. This population can grow by equal division, even without any specific distribution of cellular interiors.

Accordingly, the ribocyte must have at least 10 copies of each functional RNA and the corresponding replicase, so that chance will not regularly create new ribocytes that lack essential functions. Thus we would expect 40,000 nucleotides of RNA plus about the same amount of replicase—we are talking about a cell with less than 10^{-16} grams of RNA as its genome and ribozyme complement. Given that a modern bacterium has about a million (10^6) times as much RNA, a microbial ribocyte would have been very sparsely occupied by nucleic acids. This is true despite its relative complexity, because the calculation would yield the same conclusion even allowing for large errors. Therefore we might get a small advantage out of the dual function of RNAs as information and chemical facilitators. Ribocytes could have been exceptionally minute cells, or they could have spared interior room for unanticipated bulky functions, or perhaps both.

What might they have looked like? No one can say, save on the basis of probability. They were simpler, therefore likely small, therefore likely a spherical creature with no internal membranes and little fine structure—bag-dwellers. Only cell division of the most primitive cast would have been likely. Probably they had gene functions that could make their bounding membranes unstable, so that cells of a certain size would pinch off to release a daughter containing a sample of the parent's genes and products.

But somewhere near this point we must let them go; ribocytes and their RNAs will become clearer to us only with more study, more data, . . . and perhaps more fortunate accidents. Slow microbes they may have been, but it is hard not to feel a sneaking, possibly nepotistic, admiration for them. They made their way—and ultimately our way—decisively across a difficult time.

Readings

"Modern metabolism as a palimpsest of the RNA world." Steven A. Benner, Andrew D. Ellington, and Andreas Tauer. *Proceedings of the National Academy of Sciences of the USA* 86: 7054–7058 (1989).

How to deduce the biochemical properties of the LUCA and the ribocyte; requires some chemical reasoning.

"Relics from the RNA world." Daniel C. Jeffares, Anthony M. Poole, and David Penny. *Journal of Molecular Evolution* 46: 18–36 (1998).

An attempt to inventory the cellular functions available in an RNA world on the basis of defensible assumptions.

The Future of the RNA World

We shall never be able to study, by any method, their chemical composition or their mineralological structure.

—Auguste Comte, of the stars (1835)

I have learned to use the word "impossible" with the greatest caution.

—Wernher von Braun

Biology has at least 50 more interesting years.

—James Watson (1984)

Fun is fun and done is done.

—Stephen Edwin King

So we have come to the end. Remarkably, we have been able to produce quite a lot of evidence (and a matching volume of speculation) that life on Earth existed once, before DNA and proteins, as simple RNA microbes that had already devised the most essential of the genetic abilities we now have. Keep in sight the fact that most of the skepticism expressed toward the RNA world is directed to the role of ribocytes near the origin of life. Thinking about origins sends logic and limited knowledge to span a vast chasm. But a smaller leap can be directed more accurately. The RNA world as our immediate predecessor is adjacent in life's history on Earth, and subject to much smaller uncertainty because potent arguments using continuity and Bayes' theorem can be brought to bear on it. The contention of this book is that in the time of our immediate ancestors, RNA creatures probably owned the living Earth.

Then and after all, what is the future of this RNA world? Can we ever know how accurately we have surmised its boundaries? Can we hope to meet an RNA organism personally and take its deposition about life in RNA times? Several shadowy roads may lead toward just such a meeting.

There is the possibility that an RNA organism will surface somewhere on Earth, dragged into the light like a coelacanth—a fish once thought extinct, but now securely known to live in the sea off Madagascar. The notion of the reappearance of an RNA creature is not intrinsically outrageous; DNA microbes probably survive today from an only slightly later time, having found their way across the gigayears. As we have often observed in these pages, life has a soft but unique tenacity. Once launched on its Darwinian voyage, life can display surprising adaptive abilities, given only that the environment does not change too rapidly. For this reason, a surviving ribosaur would no doubt be found prowling a range that is ancient—unchanged or only moderately changed since the Earth's Archaean age.

Then too, we expect RNA organisms to be less capable than our fellow microorganisms. They will probably be slow to reproduce and fragile—they will not shoot out of a hot mid-ocean vent or be observed sunning themselves on a desert rock. They would also presumably reside in a place free of more modern organisms that could leave them in the dust by dividing more rapidly—or even more likely, treat them as cuisine.

Finally, ribocytes have not turned up casually, so they must at present be rare beasts. Perhaps a dedicated expedition would find them by exploring little-traveled places. This hypothetical modern province of the RNA world will have to be not only barren, untrammeled, and ancient, but also favorable to the survival of RNA creatures in other respects: perhaps it will have lower temperatures and slightly acidic waters that would act to stabilize RNA. The cold oceanic depths might be a productive hunting ground.

Somewhat less electrifying, but clearly still emitting an exciting hum, is the possibility that fossil ribocytic remains may come to light. Unfortunately, the RNA world might have been a circumscribed one—a Monaco rather than an expansive Union of Independent RNA States. However, if there once was such a region, the ribocytes may have lived and died for long enough that they drifted down to a necropolis, into which we will slice one day. It is even possible that they are hiding in the open air somewhere today, in one of the few domains of ancient, little-modified sedimentary rock at hand on the Earth's surface. Whether in a band of rock or in a core of sediment, here's hoping that, when we encounter them, we will recognize the traces of our ancestors and can say hello. Yet such a meet-and-greet will require a better knowledge than we have at present of the potential surviving chemical signature from an RNA creature.

A third possible encounter might occur on Mars. That planet was once wetter and warmer, and as it lost its atmosphere, early life might have withdrawn into one of the subsurface reservoirs of liquid water that apparently exist up to this moment. Alternatively, even if planetary change was too rapid for the ribosaurs to survive, the environmental catastrophe might have left chemical or even fossil remains that would tell us much of what we care to know. Fossils are not as explicit and detailed as a ribocyte caught on the range, but they would be good enough if the chemical residues of their nucleic acid could be identified. And with respect to our other requirements, underground Mars is probably as barren, untrammeled, and ancient as any planetary environment we can imagine. Given that prodigious explosive impacts spread early rocky material around the inner solar system, the Martian ribocytes might even be our own ancestors who, having taken an interplanetary voyage, are currently waiting for us to return—at home on a superficially frozen planet where life's clock stopped long ago. And even if Mars is barren, the solar suburbs on Europa, Enceladus, and Titan may possibly still harbor ancient lifelike materials.

Because you and I have journeyed some pleasant distance together, I will conclude with the hope that humans do stumble upon ribocytes or their ancient remains, and that I see you on that historic day. I hope we will stand shoulder to shoulder, pressing into a fluttering yellow tape, looking pensively into the pit. Perhaps a playful melody will be in the air, faintly perceptible when you do not look for it, but disappearing when you cock your head to quiet the wind. Until that fine evening, then, fare you well—may you rejoice in the tune and hear the music too.

Lexicon

In this section, some of the scientific terminology used in the text is defined in detail for easy reference by the reader.

Chromatin The complex of proteins and DNAs that makes up the bulk of chromosomes, the subcellular bearers of the genome. Chromosomes are famously seen in the cell nucleus, and pairs are evenly apportioned into daughter cells by a precise and spectacular split during cell division.

Chromosome The dye-stainable (thus chromo-), usually linear genetic bodies of creatures, composed of proteins and coiled double-stranded DNA pieces of the genome, resident in a cell's nucleus. Viral genomes are sometimes made up of RNA instead of DNA; they can be circular and single- or double-stranded and still be called chromosomes.

Continuity *or* principle of continuity A rule stating that progenitors and progeny are usually related by a small change in their genomes and therefore resemble each other. Evolutionary change (change that becomes a part of a cell's heritage) over small numbers of generations is therefore nearly continuous or smooth.

Darwinian evolution The process of descent with variation, followed by natural selection. Genetically varied progeny do not reproduce equally, necessarily yielding an increase of the frequencies of those genetic texts and genes that reproduce well.

DNA Deoxyribonucleic acid, a single- or double-stranded informational biopolymer used for genetic information storage. It consists of a linear chain of deoxyribonucleotides, phosphate-dexoyribose sugar-bases. The bases are adenine (A), guanine (G), cytosine (C), and thymine (T). *Compare* RNA.

Genome The philosophical construct that comprises all of the genes and genetic determinants of an organism, often simplified to mean the sum of all DNA or RNA sequences that make up its genes.

Group I RNAs RNAs that process themselves by removing sequences; self-splicing RNAs, hundreds to thousands of nucleotides long. These contain particular kinds of introns including a catalytic RNA (ribozyme) that removes itself from a bacterial or eucaryotic organellar RNA precursor, instead of using spliceosomes. Group I RNAs use a free G nucleotide or nucleoside cofactor to trigger self-splicing.

Group II RNAs Self-processing RNAs; self-splicing RNAs, hundreds to thousands of nucleotides long. A group II RNA is a catalytic RNA (ribozyme) within an intron that removes itself from a bacterial or eucaryotic organellar RNA precursor, instead of using spliceosomes. Group II RNAs, unlike those in group I, use no external splicing factor to initiate ribozymic intron removal.

Gya Gigayears (billions of years, multiples of 10^9 or 1,000,000,000 years) ago.

miRNA Micro RNA. *See* siRNA.

mRNA Messenger RNA, the transcript (copy) of a gene that encodes the structure of a protein. Reading from one end of the molecule to the other, thousands of nucleotides away, here is a list of the final mRNA elements: 5′ cap, 5′ untranslated region, protein information, 3′ untranslated region, with introns potentially removed throughout. The mRNA is therefore a copy of one strand of a DNA gene, capped and otherwise altered, which is sent out into the cytoplasm of cells to instruct ribosomes on the sequence of amino acids that make up a protein product (enzyme or structural protein). Messenger RNAs are highly processed or altered, in that special chemical features are added to the 5′ (caps) and 3′ (poly A) ends, and sequences (introns) are removed from the middle in order to connect up the potentially very small minority of regions (exons) that actually encode amino acid sequences in complicated organisms.

Phosphodiester bond The bond that joins nucleotides in DNA and RNA: R-O-POO$^-$-O-R, where R is ribose or ribonucleotide, P is phosphorus, and O is oxygen.

piRNA A form of small regulatory noncoding RNA somewhat paralleling miRNA and siRNA, but processed from single-stranded precursors, acting in association with its own proteins (Piwi), and restricted to control of genes and genetic invaders in animal and plant germ lines (cellular precursors of sperm and eggs).

Primer An RNA sequence of a few nucleotides, needed to replicate DNA in all chromosomes. DNA polymerases cannot initiate the synthesis of DNA but must add onto preexisting chains, usually RNA chains. Because RNA polymerases can start new chains, a special RNA polymerase (primase) makes a short RNA primer onto which DNA replication adds deoxynucleotides. These RNA primers are later removed and replaced with deoxynucleotides by a special repair mechanism that ligates the DNA replica to yield a continuous string of nucleotides.

Primordial soup The relatively concentrated natural cosmo- and geochemicals in an oceanic or pool setting that were used to assemble primordial living things.

Riboswitch A domain of hundreds of RNA nucleotides in (occasionally) an eucaryotic or (frequently) a gram-positive bacterial mRNA. These change conformation in the presence of a bound small-molecule metabolite to regulate the attached mRNA via termination of transcription or modifying the accessibility of the mRNA's ribosome binding site for translation.

RNA Ribonucleic acid, an informational biopolymer of ribonucleotides, consisting of single or doubled-and-paired strands of phosphate-ribose sugar-bases. The bases are adenine (A), guanine (G), cytosine (C), and uracil (U). *Compare* DNA.

RNAse P The RNA plus protein (ribonucleoprotein) whose catalytic RNA (hundreds of nucleotides) is the nuclease (and ribozyme) that acts in tRNA processing. By specific cleavage, RNAse creates the 5′ monophosphate end of tRNAs.

Rodinia The original supercontinent, rearranged to yield the Earth's present landmasses.

rRNA Ribosomal RNA, essential to translation in large and small versions as the framework for the ribosome. One of the large rRNAs (thousands of nucleotides) contains the peptidyl transferase and so is the catalyst for the synthesis of peptide bonds, linking together amino acids into coded proteins. Modern ribosomes are RNAs plus proteins in all cells, though protein biosynthesis probably began as an activity of pure RNA.

scnRNA Elimination of DNA sequences in the expressed nucleus of common protozoans is guided by an RNA-like mechanism. According to the "scan RNA" hypothesis, double-stranded RNAs (scnRNAs) are conveyed to a forming macronucleus, where they are used to direct elimination of particular DNA sequences by base pairing to yield the smaller, active micronucleus.

Second law of thermodynamics The principle according to which an isolated (insulated) system increases in entropy (disorder) when it undergoes spontaneous (real) change. Though this is a statistical statement, for macroscopic objects the implied decay becomes inevitable.

7S RNA A small (hundreds of nucleotides) noncoding RNA that is a structural part of the signal recognition particle (SRP). The SRP is a ribosomal accessory particle that couples the working ribosome to a receptor embedded in membranes, thereby allowing the insertion of a protein into the membrane during its simultaneous synthesis on the ribosome.

siRNA Small interfering RNA, an RNA that controls the activity of other RNAs. It acts as a negative cellular control or shutoff. Its effect on genes is based on processing double-stranded RNA from varied sources, including nuclear genes of the cell itself. Double-stranded RNAs or double-stranded RNA regions are ruler-cut at both ends into short (e.g., 21-nucleotide) duplexes with 2 nucleotide 3′ overhanging ends. The nuclease protein that does the cutting peels off and hands off one strand of this short duplex to an RNA-induced silencing complex (RISC), thereby providing its so-called Argonaut nuclease with base-pairing instructions. The siRNA guide strand finds a complementary or partially complementary sequence in an mRNA; if the complementarity is complete, a catalytic RISC will probably cut the mRNA (though it may just inhibit it), initiating its destruction and silencing the gene that it represents. *Compare* the discussion of miRNA in the next paragraph.

If the siRNA:mRNA complementarity is partial, cutting usually does not occur, but (particularly from 3′ untranslated region sites) the guide-RISC will then decrease initiation of translation of the mRNA, again silencing the gene that it represents. Small RNAs acting to decrease translation in this way are usually called micro-RNAs (miRNAs). Such miRNAs may also initiate destruction, like siRNAs. These two activities are the "classical" siRNA-miRNA silencing mechanisms; however, similar guidance by production of short regulatory RNAs incorporated into something like a RISC is much more general, and this variety is still under study. A complete account of these mechanisms will no doubt include effects on other biological events, and stimulatory as well as inhibitory actions.

snoRNA An RNA that acts during RNA modification and processing. snoRNAs occur in two styles, with characteristic subsequence blocks called H/ACA or C/D boxes. Over 200 snoRNAs of 60–150 nucleotides reside in intronic transcripts and direct enzymic modifications like pseudouridylation (H/ACA) or 2′ O-methylation (C/D) by base pairing mostly to rRNAs, but also to modified snRNA. Both kinds of snoRNAs assemble with a set of 4 specific proteins to form ribonucleoproteins, and both concentrate in the cellular nucleus in regions called the nucleolus and the Cajal bodies. Accordingly, these are sometimes called small Cajal body RNAs (scaRNAs). One such RNA is required for processing the 18S rRNA, and it is therefore essential for production of ribosomes.

snRNA An RNA processing element, one of the 5 small (100–300 nucleotides) nuclear RNAs within spliceosomes (the cellular site where noncoding internal regions, called introns, are removed from mRNA). These snRNAs, acting as parts of complex RNP particles called snRNPs (pronounced "snurps"), select the crucial boundaries of an intron by base pairing, and it is widely (but not universally) believed that they also catalyze the chemistry for precise excision of the intron and the simultaneous resealing of the bond between exons.

Telomere A special repetitive DNA sequence at the end of a chromosome, which allows the chromosome to segregate, stay separate, and be replicated. Essential sequences of the telomere are produced by reverse transcription, in which a sequence within an unusual snoRNA, which is a part of the enzyme that adds telomere DNA (telomerase), is repetitively used as a template for synthesis of a DNA telomere. This kind of RNA is of significance for human health: the genetic disease dyskeratosis congenita can be due to mutation of telomerase RNA.

tmRNA Transfer messenger RNA, a specifically bacterial RNA of hundreds of nucleotides with some characteristics of both tRNA (it becomes aminoacylated) and mRNA (it encodes a peptide). This RNA acts in a clever repair process: it inserts itself into and rescues stalled ribosomes that cannot recover because they are stalled at the end of an original mRNA that has been broken in the middle.

tRNA Transfer RNA, a small (76–92 nucleotides) translational RNA that is the repository of the genetic dictionary for encoded protein structures. It has an entry for pairing with the message codon (a 3-nucleotide anticodon) at one end, and a specific amino acid at the other end of an L-shaped tRNA structure (the latter always having the sequence . . .3′ CCA–amino acid). These aminoacyl-tRNAs plug successively into the ribosome, directed by the successive codons of an mRNA (base pairing with the tRNA anticodons), and add their amino acids to a growing protein, 5′ to 3′ on the mRNA and amino terminal end to carboxyl terminal end within the protein. The amino acid is attached to the correct Ls by protein enzymes, aminoacyl-tRNA synthetases, which probably inherited the job from RNAs that had both activities—transferring the amino acid and acting like tRNA on the ribosome. Modern data show convincingly that a tRNA-sized molecule could have carried out both functions (see Chapter 16).

Xist RNA An RNA that controls X chromosome inactivation. In mammals, Xist RNA is a long (e.g., 17,000 nucleotides in mice) transcript of the X chromatin inactivation center. This large non-coding RNA initiates inactivation of more than one X chromosome in females (to make gene expression in XY males and XX females the same). X inactivation leads to histone and DNA modifications, and it may be controlled by double-stranded RNA that results from Xist pairing to Tsix, a complementary overlapping X chromosomal RNA. Repetitive sequence elements like centromeres, telomeres, and some genes are also modified under similar double-stranded RNA control. Small RNAs and RNA-induced silencing complex also operate in somewhat parallel control of selfish RNAs that can move within genomes. *Compare* siRNA.

Index